Country of Origin: United States

Springer

the language of science

T0076683

Thank you for your order! An invoice is being mailed separately. Pl
For more information on our products and for new book annou

Reference# 0081369539

Ship to:

BAKER & TAYLOR BOOKS
251 MT OLIVE CHURCH ROAD
COMMERCE SERVICE CENTER

Tel: 8154722444

COMMERCE GA 305991100 US

US

Quantity	ISBN	Title
1	9781461422983	Graphical Models with R

1 Total quantity enclosed

Claim Policy:

Claims regarding this shipment must be confirmed in writing and received by us within thirty days of the invoice date. Please call our toll-free number if assistance is required.

Book Return Policy

Returns are accepted within two months of invoice date. Returns of products that were purchased for resale are accepted within twelve months of invoice date. Please enclose a copy of our invoice or packing slip when you send your return to expedite processing and avoid a possible 10% penalty. For your protection, returns should be sent by a traceable method. We are not responsible for product returned that does not belong to us.

0081369539

Packing List

20121227

PICKUP   B & T UPS Collect

Purchase Order: COM1353206

	Author/Editor
	Højsgaard, Søren

Use R!

Series Editors:
Robert Gentleman   Kurt Hornik   Giovanni G. Parmigiani

For further volumes:
www.springer.com/series/6991

Søren Højsgaard · David Edwards ·
Steffen Lauritzen

Graphical Models with R

Søren Højsgaard
Department of Mathematical Sciences
Aalborg University
Aalborg
Denmark

David Edwards
Centre for Quantitative Genetics and
Genomics
Department of Molecular Biology and
Genetics
Aarhus University
Aarhus
Denmark

Steffen Lauritzen
Department of Statistics
University of Oxford
Oxford
UK

Series Editors:
Robert Gentleman
Program in Computational Biology
Division of Public Health Sciences
Fred Hutchinson Cancer Research Center
Seattle, WA
USA

Giovanni G. Parmigiani
The Sidney Kimmel Comprehensive
Cancer Center at Johns Hopkins University
Baltimore, MD
USA

Kurt Hornik
Department of Statistik and Mathematik
Wirtschaftsuniversität Wien
Wien
Austria

ISBN 978-1-4614-2298-3   e-ISBN 978-1-4614-2299-0
DOI 10.1007/978-1-4614-2299-0
Springer New York Dordrecht Heidelberg London

Library of Congress Control Number: 2012931941

Printed on acid-free paper

Springer is part of Springer Science+Business Media (www.springer.com)

Preface

Graphical models in their modern form have been around since the late 1970s and appear today in many areas of the sciences. Along with the ongoing developments of graphical models, a number of different graphical modelling software programs have been written over the years. In recent years many of these software developments have taken place within the R community, either in the form of providing an R interface to existing software or in the form of new R packages. At the time of writing, the taskview for graphical models in R at

http://cran.r-project.org/web/views/gR.html

lists some thirty packages related to graphical models. It is expected that this number will grow considerably, and the packages will be extended and modified.

This book attempts to give the reader a gentle introduction to graphical modelling using R and the main features of some of these packages, hopefully sharpening the appetite for using and developing these packages further. In addition, we shall give a few examples of how more advanced aspects of graphical modelling can be represented and handled within R.

We emphasize that this book is not a manual to the collection of packages mentioned and the general theory of the models is only described to an extent which allows the book to be read meaningfully on its own. For a more extensive description of the theory we refer to the textbooks available, such as Whittaker (1990), Lauritzen (1996), and Edwards (2000).

The organization of the book is as follows:

Chapter 1 treats graphs without any direct reference to statistical models although the significance of graphs for conditional independence is briefly explained and exemplified. This chapter may be skipped at first reading and returned to as needed.

Chapter 2 discusses graphical models for contingency tables, i.e. graphical models for discrete data. Chapter 3 deals with Bayesian networks and the updating of conditional probabilities. Chapter 4 deals with graphical models for the normal distribution, i.e. for continuous data. Chapter 5 discusses mixed interaction models which refers to a combination of discrete and continuous variables and this chapter thus unifies Chaps. 2 and 4.

Chapters 2, 4 and 5 all deal with models which are largely data-driven, mostly analysed within a frequentist perspective; these chapters constitute the core of the book.

Chapter 6 discusses graphical models for complex stochastic systems with focus on methods of inference which involve Markov chain Monte Carlo sampling (Gilks et al. 1994). Both Chaps. 3 and 6 deal with models which strongly exploit prior substantive knowledge and are mostly treated within a Bayesian perspective.

A perspective on graphical models which has become particular important in the last decades involves their ability to deal with problems involving data of high dimension. This aspect is dealt with in Chap. 7.

Finally, we would to thank Sofia Massa, Clive Bowsher and Vanessa Didelez for reading early drafts of this book and providing us with encouragement and constructive comments.

Aalborg, Denmark   Søren Højsgaard
Tjele, Denmark   David Edwards
Oxford, UK   Steffen Lauritzen

Contents

Chapter 1
Graphs and Conditional Independence

1.1 Introduction

A graph as a *mathematical* object may be defined as a pair $\mathcal{G} = (V, E)$, where V is a set of *vertices* or *nodes* and E is a set of *edges*. Each edge is associated with a pair of nodes, its *endpoints*. Edges may in general be directed, undirected, or bidirected. Graphs are typically visualized by representing nodes by circles or points, and edges by lines, arrows, or bidirected arrows. We use the notation $\alpha - \beta$, $\alpha \rightarrow \beta$, and $\alpha \leftrightarrow \beta$ to denote edges between α and β. Graphs are useful in a variety of applications, and a number of packages for working with graphs are available in R.

We have found the **graph** package to be particularly useful, since it provides a way of representing graphs as so-called `graphNEL` objects (graphs as *N*ode and *E*dge *L*ists) and thereby gives access to a wide range of graph-theoretic operations (in the **graph** package), efficient implementations of standard graph algorithms (in the **RBGL** package), and allows the graphs to be easily displayed in a variety of layouts (using the **Rgraphviz** package from BioConductor). Much of this book uses this representation. In statistical applications we are particularly interested in two special graph types: undirected graphs and directed acyclic graphs (often called DAGs).

We have also found the **igraph** package to be useful. Like the **graph** package, **igraph** supports both undirected and directed graphs and implements various graph algorithms. Functions in the package allow graphs to be displayed in a variety of formats. The internal representation of graphs in the **igraph** package differs from the representation in the **graph** package.

The **gRbase** package supplements **graph** and **igraph** by implementing some algorithms useful in graphical modelling. **gRbase** also provides two wrapper functions, `ug()` and `dag()`, for easily creating undirected graphs and DAGs represented either as `graphNEL` objects (the default), `igraph` objects or adjacency matrices.

The first sections of this chapter describe some of the most useful functions available when working with graphical models. These come variously from the **gRbase**, **graph** and **RBGL** packages, but it is not usually necessary to know which. To use the functions and plot the graphs it is enough to load **gRbase** and **Rgraphviz**, since

S. Højsgaard et al., *Graphical Models with R*, Use R!,
DOI 10.1007/978-1-4614-2299-0_1, © Springer Science+Business Media, LLC 2012

gRbase automatically loads the other packages. To find out which package a function we mention belongs to, please refer to the index to this book or to the usual help functions.

As *statistical* objects, graphs are used to represent models, with nodes representing model variables (and sometimes model parameters) in such a way that the independence structure of the model can be read directly off the graph. Accordingly, a section of this chapter is devoted to a brief description of the key concept of conditional independence and explains how this is linked to graphs. Throughout the book we shall repeatedly return to this in more detail.

1.2 Graphs

Our graphs have a finite node set V and for the most part they are *simple graphs* in the sense that they have no loops nor multiple edges. Two vertices α and β are said to be *adjacent*, written $\alpha \sim \beta$, if there is an edge between α and β in $\mathcal{G}$, i.e. if either $\alpha - \beta$, $\alpha \to \beta$, or $\alpha \leftrightarrow \beta$.

In this chapter we primarily represent graphs as graphNEL objects, and except where stated otherwise, the functions we describe operate on these objects. However, different representations can be useful for various reasons, for example igraph objects or as adjacency matrices. It is easy to convert between graphNEL objects, igraph objects and adjacency matrices, using the as() function. For example, if gNg is a graphNEL object, then

```
> ig <- as(gNg, "igraph")
> ag <- as(gNg, "matrix")
```

creates versions of the same graph represented as an igraph object and as an adjacency matrix. Similarly, to convert back we could write

```
> gNg <- as(ig, "graphNEL")
```

1.2.1 Undirected Graphs

An undirected graph may be created using the ug() function. The graph can be specified using a list of formulas, a single formula or a list of vectors. Thus the following forms are equivalent:

```
> library(gRbase)
> ug0 <- ug(~a:b, ~b:c:d, ~e)
> ug0 <- ug(~a:b+b:c:d+e)
> ug0 <- ug(~a*b+b*c*d+e)
> ug0 <- ug(c("a","b"),c("b","c","d"),"e")
> ug0

A graphNEL graph with undirected edges
Number of Nodes = 5
Number of Edges = 4
```

graphNEL graphs are displayed with `plot()`:

```
> library(Rgraphviz)
> plot(ug0)
```

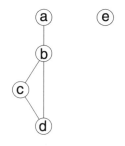

Per default the ug() function returns an graphNEL object, but the options result="igraph" or result="matrix" lead it to return an igraph or adjacency matrix instead. For example,

```
> ug0i <- ug(~a:b+b:c:d+e, result="igraph")
> ug0i
```

```
Vertices: 5
Edges: 4
Directed: FALSE
Edges:

[0] 'a' -- 'b'
[1] 'b' -- 'c'
[2] 'b' -- 'd'
[3] 'c' -- 'd'
```

There is a plot() method for igraph objects in the **igraph** package. There are also various facilities for controlling the layout. For example, we may use a layout algorithm called layout.spring as follows:

```
> plot(ug0i, layout=layout.spring)
```

The default size of vertices and their labels is quite small. This is easily changed by setting certain attributes on the graph, see Sect. 1.4.3 for examples. However, to avoid changing these attributes for all the graphs shown in the following we have defined a small plot function myiplot() as follows:

```
> myiplot <- function(x, ...){
+     V(x)$size <- 30
+     V(x)$label.cex <- 3
+     plot(x,...)
+ }
```

The graph ug0i is then displayed with:

```
> myiplot(ug0i, layout=layout.spring)
```

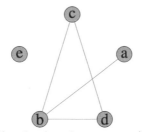

Edges can be added and deleted using the addEdge() and removeEdge() functions:

```
> ug0a <- addEdge("a","c", ug0)
> ug0a <- removeEdge("c","d", ug0)
```

The nodes and edges of a graph can be retrieved with nodes() and edges() functions.

```
> nodes(ug0)

[1] "a" "b" "c" "d" "e"

> edges(ug0)

$a
[1] "b"

$b
[1] "c" "d" "a"

$c
[1] "d" "b"

$d
[1] "b" "c"

$e
character(0)
```

Thus edges() gives, for each node, the adjacent nodes. The edgeList() function, that returns a list of (unordered) pairs, may be used (to compact the output, we prefix the command with function str()).

```
> str(edgeList(ug0))

List of 4
 $ : chr [1:2] "b" "a"
 $ : chr [1:2] "c" "b"
 $ : chr [1:2] "d" "b"
 $ : chr [1:2] "d" "c"
```

A subset $A \subseteq V$ is *complete* if all vertex pairs in A are connected by an edge. A graph $\mathcal{G} = (V, E)$ is complete if the vertex set V is complete. A *clique* is a maximal complete subset, that is to say, a complete subset that is not contained in a larger complete subset. The set of cliques of a graph $\mathcal{G}$ is denoted by $\mathcal{C}(\mathcal{G})$. Note that in the literature the term clique is often used to denote a complete subset and may not necessarily be maximal. The function maxClique() returns the (maximal) cliques of a graph:

```
> is.complete(ug0)
```

[1] FALSE

```
> is.complete(ug0, c("b","c","d"))
```

[1] TRUE

```
> maxClique(ug0)
```

```
$maxCliques
$maxCliques[[1]]
[1] "b" "c" "d"

$maxCliques[[2]]
[1] "b" "a"

$maxCliques[[3]]
[1] "e"
```

A *path* (of length n) between α and β in an undirected graph is a set of vertices $\alpha = \alpha_0, \alpha_1, \ldots, \alpha_n = \beta$ where $\alpha_{i-1} - \alpha_i$ for $i = 1, \ldots, n$. If a path $\alpha = \alpha_0, \alpha_1, \ldots, \alpha_n = \beta$ has $\alpha = \beta$ then the path is said to be a *cycle* of length n.

A subset $D \subset V$ in an undirected graph is said to *separate* $A \subset V$ from $B \subset V$ if every path between a vertex in A and a vertex in B contains a vertex from D.

```
> separates("a", "d", c("b", "c"), ug0)
```

[1] TRUE

This shows that $\{b, c\}$ separates $\{a\}$ and $\{d\}$.

The graph $\mathcal{G}_0 = (V_0, E_0)$ is said to be a *subgraph* of $\mathcal{G} = (V, E)$ if $V_0 \subseteq V$ and $E_0 \subseteq E$. For $A \subseteq V$, let E_A denote the set of edges in E between vertices in A. Then $\mathcal{G}_A = (A, E_A)$ is the *subgraph induced by* A. For example

```
> ug1 <- subGraph(c("b","c","d","e"), ug0)
> plot(ug1)
```

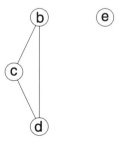

The *boundary* $\text{bd}(\alpha) = \text{adj}(\alpha)$ is the set of vertices adjacent to α and for undirected graphs the boundary is equal to the set of *neighbours* $\text{ne}(\alpha)$. The *closure* $\text{cl}(\alpha)$ is $\text{bd}(\alpha) \cup \{\alpha\}$.

```
> adj(ug0, "c")

$c
[1] "d" "b"
> closure("c", ug0)

     c1   c2
"c" "d" "b"
```

1.2.2 Directed Acyclic Graphs

A *directed graph* as a mathematical object is a pair $\mathcal{G} = (V, E)$ where V is a set of vertices and E is a set of directed edges, normally drawn as arrows. A directed graph is *acyclic* if it has no directed cycles, that is, cycles with the arrows pointing in the same direction all the way around. A *DAG* is a directed graph that is acyclic.

A DAG may be created using the dag() function. The graph can be specified by a list of formulas or by a list of vectors. The following statements are equivalent:

```
> dag0 <- dag(~a, ~b*a,  ~c*a*b, ~d*c*e, ~e*a, ~g*f)
> dag0 <- dag(~a + b*a + c*a*b + d*c*e + e*a + g*f)
> dag0 <- dag(~a + b|a + c|a*b + d|c*e + e|a + g|f)
> dag0 <- dag("a", c("b","a"), c("c","a","b"), c("d","c","e"),
+             c("e","a"),c("g","f"))
> dag0
A graphNEL graph with directed edges
Number of Nodes = 7
Number of Edges = 7
```

Note that ~a means that "a" has no parents while ~d*b*c means that "d" has parents "b" and "c". Instead of "*", a ":" can be used in the specification. If the specified graph contains cycles then dag() returns NULL.

Per default the dag() function returns a graphNEL object, but the options result="igraph" or result="matrix" lead it to return an igraph or adjacency matrix instead.

DAGs are displayed with plot():

```
> plot(dag0)
```

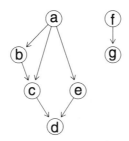

The nodes and edges of a DAG can be retrieved with the nodes() and edges() functions.

```
> nodes(dag0)
```

```
[1] "a" "b" "c" "d" "e" "g" "f"
```

```
> str(edges(dag0))
```

```
List of 7
 $ a: chr [1:3] "b" "c" "e"
 $ b: chr "c"
 $ c: chr "d"
 $ d: chr(0)
 $ e: chr "d"
 $ g: chr(0)
 $ f: chr "g"
```

Thus edges() gives the children of each node. Alternatively a list of (ordered) pairs can be optained with edgeList()

```
> str(edgeList(dag0))
```

```
List of 7
 $ : chr [1:2] "a" "b"
 $ : chr [1:2] "a" "c"
 $ : chr [1:2] "b" "c"
 $ : chr [1:2] "c" "d"
 $ : chr [1:2] "e" "d"
 $ : chr [1:2] "a" "e"
 $ : chr [1:2] "f" "g"
```

The vpar() function returns a list, with an element for each node together with its parents:

```
> vpardag0 <- vpar(dag0)
> vpardag0$c
```

```
[1] "c" "a" "b"
```

A *path* (of length n) from α to β is a sequence of vertices $\alpha = \alpha_0, \ldots, \alpha_n = \beta$ such that $\alpha_{i-1} \to \alpha_i$ is an edge in the graph. If there is a path from α to β we write $\alpha \mapsto \beta$. The *parents* pa(β) of a node β are those nodes α for which $\alpha \to \beta$. The *children* ch(α) of a node α are those nodes β for which $\alpha \to \beta$. The *ancestors* an(β) of a node β are the nodes α such that $\alpha \mapsto \beta$. The *ancestral set* an(A) of a set A is the union of A with its ancestors. The *ancestral graph* of a set A is the subgraph induced by the ancestral set of A.

```
> parents("d",dag0)
```

```
[1] "c" "e"
> children("c",dag0)
```

```
[1] "d"
> ancestralSet(c("b","e"),dag0)
```

```
[1] "a" "b" "e"
> ancestralGraph(c("b","e"),dag0)
```

```
A graphNEL graph with directed edges
Number of Nodes = 3
Number of Edges = 2
```

```
> plot(ancestralGraph(c("b","e"),dag0))
```

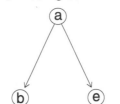

An important operation on DAGs is to (i) add edges between the parents of each node, and then (ii) replace all directed edges with undirected ones, thus returning an undirected graph. This operation is used in connection with independence interpretations of the DAG, see Sect. 1.3, and is known as *moralization*. This is implemented by the `moralize()` function:

```
> dag0m <- moralize(dag0)
```

```
A graphNEL graph with undirected edges
Number of Nodes = 7
Number of Edges = 8
```

```
> plot(dag0m)
```

```
[1] "A graph with 7 nodes."
```

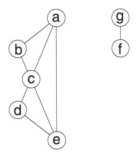

1.2.3 Mixed Graphs

Although the primary focus of this book is on undirected graphs and DAGs, it is also useful to consider *mixed graphs*. These are graphs with at least two types of edges, for example directed and undirected, or directed and bidirected.

A sequence of vertices $v_1, v_2, \ldots, v_k, v_{k+1}$ is called a *path* if for each $i = 1, \ldots, k$, either $v_i - v_{i+1}$, $v_i \leftrightarrow v_{i+1}$ or $v_i \rightarrow v_{i+1}$. If $v_i - v_{i+1}$ for each i the path is called *undirected*, if $v_i \rightarrow v_{i+1}$ for each i it is called *directed*, and if $v_i \rightarrow v_{i+1}$ for at least one i it is called *semi-directed*. If $v_i = v_{k+1}$ it is called a *cycle*.

Mixed graphs are represented in both the **graph** and **igraph** packages as directed graphs with multiple edges. In this sense they are not simple. A convenient way of defining them (in lieu of model formulae) is to use adjacency matrices. We can construct such a matrix as follows:

```
> adjm <- matrix(c(0,1,1,0,1,0,0,1,1,0,0,0,1,1,1,0), nrow=4)
> rownames(adjm) <- colnames(adjm) <- letters[1:4]
> adjm

  a b c d
a 0 1 1 1
b 1 0 0 1
c 1 0 0 1
d 0 1 0 0
```

We can use this matrix to create a graphNEL object

```
> gG <- as(adjm, "graphNEL")
> plot(gG, "neato")
```

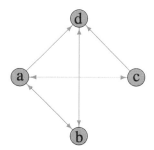

Note that **Rgraphviz** interprets symmetric entries as double-headed arrows and thus does not distinguish between bidirected and undirected edges. The same is true if we display the graph as a igraph object:

```
> gG1 <- as(adjm, "igraph")
> myiplot(gG1, layout=layout.spring)
```

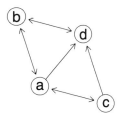

However we can persuade **igraph** to display undirected instead of bidirected edges, as follows:

```
> E(gG1)$arrow.mode <- c(2,0)[1+is.mutual(gG1)]
> myiplot(gG1, layout=layout.spring)
```

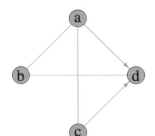

To do this using **Rgraphviz** is rather more complex: we give an example later in Sect. 1.4.2.

A *chain graph* is a mixed graph with no bidirected edges and no semi-directed cycles. Such graphs form a natural generalisation of undirected graphs and DAGs, as we shall see later. The following example is from Frydenberg (1990a):

```
> d1 <- matrix(0,11,11)
> d1[1,2] <- d1[2,1] <- d1[1,3] <- d1[3,1] <- d1[2,4] <- d1[4,2] <-
+     d1[5,6] <- d1[6,5] <- 1
> d1[9,10] <- d1[10,9] <- d1[7,8] <- d1[8,7] <- d1[3,5] <-
+     d1[5,10] <- d1[4,6] <- d1[4,7] <- 1
> d1[6,11] <- d1[7,11] <- 1
> rownames(d1) <- colnames(d1) <- letters[1:11]
> cG1 <- as(d1, "igraph")
> E(cG1)$arrow.mode <- c(2,0)[1+is.mutual(cG1)]
> myiplot(cG1, layout=layout.spring)
```

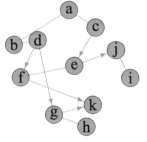

The *components* of a chain graph $\mathcal{G}$ are the connected components of the graph formed after removing all directed edges from $\mathcal{G}$. All edges within a component are undirected, and all edges between components are directed. Also, all arrows between any two components have the same direction. The graph constructed by identifying its nodes with the components of $\mathcal{G}$, and joining two nodes with an arrow whenever there is an arrow between the corresponding components in $\mathcal{G}$, is a DAG, the so-called *component DAG* of $\mathcal{G}$, written $\mathcal{G}_C$.

The is.chaingraph() function in the **lcd** package determines whether a mixed graph is a chain graph. It takes an adjacency matrix as input. For example, the above graph is indeed a chain graph:

```
> library(lcd)
> is.chaingraph(as(cG1, "matrix"))
```

```
$result
[1] TRUE

$vert.order
[1]  1 2 3 4 5 6 7 8 9 10 11

$chain.size
[1] 4 2 2 2 1
```

Here vert.order gives an ordering of the vertices, from which the connected components may be identified using chain.size.

The *anterior set* of a vertex set $S \subseteq V$ is defined in terms of the component DAG. Write the set of components of $\mathcal{G}$ containing S as S_c. Then the anterior set of S in $\mathcal{G}$ is defined as the union of the components in the ancestral set of S_c in $\mathcal{G}_C$. The *anterior graph* of $S \subseteq V$ is the subgraph of $\mathcal{G}$ induced by the anterior set of S.

The *moralization* operation is also important for chain graphs. Similar to DAGs, unmarried parents of the same chain components are joined and directions are then removed. The operation is implemented in the moralize() function in the **lcd** package, which uses the adjacency matrix representation. For example,

```
> cGm <- as(moralize(as(cG1, "matrix")), "graphNEL")
> plot(cGm)
```

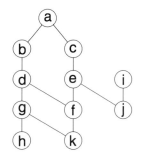

1.3 Conditional Independence and Graphs

The concept of statistical independence is presumably familiar to all readers but that of *conditional independence* may be less so. Suppose that we have a collection of random variables $(X_v)_{v \in V}$ with a joint density. Let A, B and C be subsets of V and let $X_A = (X_v)_{v \in A}$ and similarly for X_B and X_C. Then the statement that X_A and X_B are conditionally independent given X_C, written $A \perp\!\!\!\perp B \mid C$, means that for each possible value of x_C of X_C, X_A and X_B are independent in the conditional distribution given $X_C = x_c$. So if we write $f()$ for a generic density or probability mass function, then one characterization of $A \perp\!\!\!\perp B \mid C$ is that

$$f(x_A, x_B \mid x_C) = f(x_A \mid x_C) f(x_B \mid x_C).$$

An equivalent characterization (Dawid 1998) is that the joint density of (X_A, X_B, X_C) factorizes as

$$f(x_A, x_B, x_C) = g(x_A, x_C) h(x_B, x_C), \tag{1.1}$$

that is, as a product of two functions $g()$ and $h()$, where $g()$ does not depend on x_B and $h()$ does not depend on x_A. This is known as the *factorization criterion*.

Parametric models for $(X_v)_{v \in V}$ may be thought of as specifying a set of joint densities (one for each admissible set of parameters). These may admit factorisations of the form just described, giving rise to conditional independence relations between the variables. Some models give rise to patterns of conditional independences that can be represented as an undirected graph. More specifically, let $\mathcal{G} = (V, E)$ be an undirected graph with cliques $C_1, \ldots, C_k$. Consider a joint density $f()$ of the variables in V. If this admits a factorization of the form

$$f(x_V) = \prod_{i=1}^{k} g_i(x_{C_i})$$

for some functions $g_1() \ldots g_k()$ where $g_j()$ depends on x only through x_{C_j} then we say that $f()$ factorizes according to $\mathcal{G}$.

If all the densities under a model factorize according to $\mathcal{G}$, then $\mathcal{G}$ encodes the conditional independence structure of the model, through the following result (the *global Markov property*): whenever sets A and B are separated by a set C in the graph, then $A \perp\!\!\!\perp B \mid C$ under the model. Thus for example

```
> plot(ug0)
```

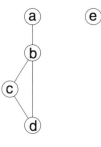

```
> separates("a", "d", "b", ug0)
```

```
[1] TRUE
```

shows that under a model with this dependence graph, $a \perp\!\!\!\perp d \mid b$.

If we want to find out whether two variable sets are marginally independent, we ask whether they are separated by the empty set, which we specify using a character vector of length zero:

```
> separates("a", "d", character(0), ug0)
```

```
[1] FALSE
```

Model families that admit suitable factorizations are described in later chapters in this book. These include: log-linear models for multivariate discrete data, graphical Gaussian models for multivariate Gaussian data, and mixed interaction models for mixed discrete and continuous data.

Other models give rise to patterns of conditional independences that can be represented by DAGs. These are models for which the variable set V may be ordered in such way that the joint density factorizes as follows

$$f(x_V) = \prod_{v \in V} f(x_v \mid x_{\text{pa}(v)}) \tag{1.2}$$

for some variable sets $\{\text{pa}(v)\}_{v \in V}$ such that the variables in $\text{pa}(v)$ precede v in the ordering. Again the vertices of the graph represent the random variables, and we can identify the sets $\text{pa}(v)$ with the parents of v in the DAG.

With DAGs, conditional independence is represented by a property called *d-separation*. That is, whenever sets A and B are d-separated by a set C in the graph, then $A \perp\!\!\!\perp B \mid C$ under the model. The notion of d-separation can be defined in various ways, but one characterisation is as follows: A and B are d-separated by a set C if and only if they are separated in the graph formed by moralizing the anterior graph of $A \cup B \cup C$.

So we can easily define a function to test this:

```
> d.separates <- function(a,b,c,dag) {
+     separates(a,b,c,
+              gRbase::moralize(ancestralGraph(union(union(a,b),c),
               dag)))}
> d.separates("c", "e", "a", dag0)

[1] TRUE
```

So under dag0 it holds that $c \perp\!\!\!\perp e \mid a$.

Alternatively, we can use the function dSep() in the **ggm** package:

```
> library(ggm)
> dSep(as(dag0, "matrix"), "c", "e", "a")

[1] TRUE
```

Still other models correspond to patterns of conditional independences that can be represented by a chain graph $\mathcal{G}$. There are several ways to relate Markov properties to chain graphs. Here we describe the so-called LWF Markov properties, associated with Lauritzen, Wermuth and Frydenberg.

For these there are two levels to the factorization requirements. Firstly, the joint density needs to factorize in a way similar to a DAG, i.e.

$$f(x_V) = \prod_{C \in \mathcal{C}} f(x_C \mid x_{\text{pa}(C)})$$

where $\mathcal{C}$ is the set of components of $\mathcal{G}$. In addition, each conditional density $f(x_C \mid x_{\text{pa}(C)})$ must factorize according to an undirected graph constructed in the following way. First form the subgraph of $\mathcal{G}$ induced by $C \cup \text{pa}(C)$, drop directions, and then complete $\text{pa}(C)$ (that is, add edges between all vertices in $\text{pa}(C)$).

For densities which factorize as above, conditional independence is related to a property called *c-separation*: that is, $A \perp\!\!\!\perp B \mid C$ whenever sets A and B are c-separated by C in the graph. The notion of c separation in chain graphs is similar to that of d-separation in DAGs. A and B are c-separated by a set C if and only if they are separated in the graph formed by moralizing the anterior graph of $A \cup B \cup C$. The is.separated() function in the **lcd** package can be used to query a given chain graph for c-separation. For example,

```
> library(lcd)
> is.separated("e", "g", c("k"), as(cG1,"matrix"))
```

```
[1] FALSE
```

implies that $e \not\!\perp\!\!\!\perp g \mid k$ for the chain graph cG1 we considered previously.

1.4 More About Graphs

1.4.1 Special Properties

A node in an undirected graph is *simplicial* if its boundary is complete.

```
> is.simplicial("b", ug0)
```

```
[1] FALSE
```

```
> simplicialNodes(ug0)
```

```
[1] "a" "c" "d" "e"
```

To obtain the *connected components* of a graph:

```
> connComp(ug0)
```

```
[[1]]
[1] "a" "b" "c" "d"

[[2]]
[1] "e"
```

If a cycle $\alpha = \alpha_0, \alpha_1, \ldots, \alpha_n = \alpha$ has adjacent elements $\alpha_i \sim \alpha_j$ with $j \notin \{i-1, i+1\}$ then it is said to have a *chord*. If it has no chords it is said to be *chordless*. A graph with no chordless cycles of length ≥ 4 is called *triangulated* or *chordal*:

```
> is.triangulated(ug0)
```

```
[1] TRUE
```

Triangulated graphs are of special interest for graphical models as they admit closed-form maximum likelihood estimates and allow considerable computational simplification by decomposition.

A triple (A, B, D) of non-empty disjoint subsets of V is said to *decompose* $\mathcal{G}$ into $\mathcal{G}_{A \cup D}$ and $\mathcal{G}_{B \cup D}$ if $V = A \cup B \cup D$ where D is complete and separates A and B.

```
> is.decomposition("a", "d", c("b","c"), ug0)
```

```
[1] FALSE
```

Note that although $\{d\}$ is complete and separates $\{a\}$ and $\{b, c\}$ in ug0, the condition fails because $V \neq \{a, b, c, d\}$.

A graph is *decomposable* if it is complete or if it can be decomposed into decomposable subgraphs. A graph is decomposable if and only if it is triangulated.

An ordering of the nodes in a graph is called a *perfect ordering* if bd(i) ∩ $\{1, \dots, i-1\}$ is complete for all i. Such an ordering exists if and only if the graph is triangulated. If the graph is triangulated, then a perfect ordering can be obtained with the *maximum cardinality search* (or mcs) algorithm. The mcs() function will produce such an ordering if the graph is triangulated; otherwise it will return NULL.

```
> mcs(ug0)

[1] "a" "b" "c" "d" "e"
```

Sometimes it is convenient to have some control over the ordering given to the variables:

```
> mcs(ug0, root=c("d","c","a"))

[1] "d" "c" "b" "a" "e"
```

Here mcs() tries to follow the ordering given and succeeds for the first two variables but then fails afterwards.

The cliques of a triangulated undirected graph can be ordered as $(C_1, \dots, C_Q)$ to have the *running intersection property* (also called a *RIP ordering*). The running intersection property is that $C_j \cap (C_1 \cup \cdots \cup C_{j-1}) \subset C_i$ for some $i < j$ for $j = 2, \dots, Q$. We define the sets $S_j = C_j \cap (C_1 \cup \cdots \cup C_{j-1})$ and $R_j = C_j \setminus S_j$ with $S_1 = \emptyset$. The sets S_j are called *separators* as they separate R_j from $(C_1 \cup \cdots \cup C_{j-1}) \setminus S_j$. Any clique C_i where $S_j \subset C_i$ with $i < j$ is a possible parent of C_j. The rip() function returns such an ordering if the graph is triangulated (otherwise, it returns list()):

```
> rip(ug0)

cliques
   1 : b a
   2 : d b c
   3 : e
separators
   1 :
   2 : b
   3 :
parents
   1 : 0
   2 : 1
   3 : 0
```

If a graph is not triangulated it can be made so by adding extra edges, so called *fill-ins*, using triangulate():

```
> ug2 <- ug(~a:b:c+c:d+d:e+a:e)
> is.triangulated(ug2)

[1] FALSE

> plot(ug2)
```

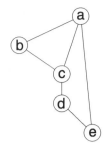

```
> ug3 <- triangulate(ug2)
> is.triangulated(ug3)
```

`[1] TRUE`

```
> plot(ug3)
```

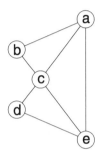

Recall that an undirected graph $\mathcal{G}$ is triangulated (or chordal) if it has no cycles of length $>= 4$ without a chord. A graph is triangulated if and only if there exists a perfect ordering of its vertices. Any undirected graph $\mathcal{G}$ can be triangulated by adding edges to the graph, so called fill-ins, resulting in a graph $\mathcal{G}^*$, say. Some of the fill-ins on $\mathcal{G}^*$ may be superfluous in the sense that they could be removed and still give a triangulated graph. A triangulation with no superfluous fill-ins is called a *minimal* triangulation. In general this is not unique. This should be distinguished from a *minimum* triangulation which is a graph with the smallest number of fill-ins. Finding a minimum triangulation is known to be NP-hard. The function `minimalTriang()` finds a minimal triangulation. Consider the following:

```
> G1 <- ug(~a:b+b:c+c:d+d:e+e:f+a:f+b:e)
> mt1.G1 <- minimalTriang(G1)
> G2 <- ug(~a:b:e:f+b:c:d:e)
> mt2.G1<-minimalTriang(G1, TuG=G2)
> par(mfrow=c(2,2))
> plot(G1, sub="G1")
> plot(mt1.G1, sub="mt1.G1")
> plot(G2, sub="G2")
> plot(mt2.G1, sub="mt2.G1")
```

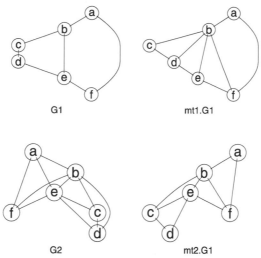

The graph G1 is not triangulated; mt1.G1 is a minimal triangulation of G1. Furthermore, G2 is a triangulation of G1, but it is not a minimal triangulation. Finally, mt2.G1 is a minimal triangulation of G1 formed on the basis of G2.

The *maximal prime subgraph decomposition* of an undirected graph is the smallest subgraphs into which the graph can be decomposed. Consider the following code fragment:

```
> G1 <- ug(~a:b+b:c+c:d+d:e+e:f+a:f+b:e)
> G1.rip <- mpd(G1)
> G1.rip

cliques
  1 : f a b e
  2 : d b e c
separators
  1 :
  2 : b e
parents
  1 : 0
  2 : 1

> par(mfrow=c(1,3))
> plot(G1, main="G1")
> plot(subGraph(G1.rip$cliques[[1]], G1), main="subgraph 1")
> plot(subGraph(G1.rip$cliques[[2]], G1), main="subgraph 2")


cliques
  1 : f a b e
  2 : d b e c
separators
  1 :
  2 : b e
parents
  1 : 0
  2 : 1
```

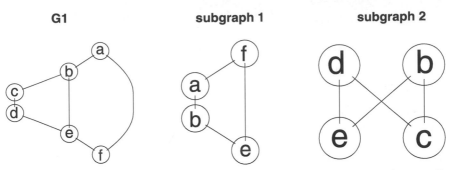

G1   **subgraph 1**   **subgraph 2**

Here G1 is not decomposable but the graph can be decomposed. The function mpd()
returns a junction RIP-order representation of the maximal prime subgraph decomposition. The subgraphs of G1 defined by the cliques listed in G1.rip are the smallest subgraphs into which G1 can be decomposed.

The *Markov blanket* of a vertex v in a DAG $\mathcal{G}$ may be defined as the minimal
set that d-separates v from the remaining variables. It is easily derived as the set of
neighbours to v in the moral graph of $\mathcal{G}$. For example, the Markov blanket of vertex
e in dag0 is

```
> adj(moralize(dag0), "e")
```

```
$e
[1] "a" "c" "d"
```

It is easily seen that the Markov blanket of v is the union of v's parents, v's children,
and the parents of v's children.

1.4.2 *Graph Layout in* **Rgraphviz**

Although the way graphs are displayed on the page or screen has no bearing on their
mathematical or statistical properties, in practice it is helpful to display them in a
way that clearly reveals their structure. The **Rgraphviz** package implements several
methods for automatically setting graph layouts. We sketch these very briefly here:
for more detailed information see the online help files, for example, type ?dot.

- The *dot* method, which is default, is intended for drawing DAGs or hierarchies
  such as organograms or phylogenies.
- The *twopi* method is suitable for connected graphs: it produces a circular layout
  with one node placed at the centre and others placed on a series of concentric
  circles about the centre.
- The *circo* method also produces a circular layout.
- The *neato* method is suitable for undirected graphs: an iterative algorithm determines the coordinates of the nodes so that the geometric distance between node-
  pairs approximates their path distance in the graph.
- Similarly, the *fdp* method is based on an iterative algorithm due to Fruchterman
  and Reingold (1991), in which adjacent nodes are attracted and non-adjacent
  nodes are repulsed.

The graphs displayed using `Rgraphviz` can also be embellished in various ways: the following example displays the text in red and fills the nodes with light grey.

```
> plot(dag0, attrs=list(node = list(fillcolor="lightgrey",
       fontcolor="red")))
```

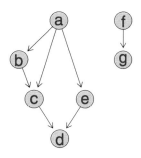

Graph layouts can be reused: this can be useful, for example to would-be authors of books on graphical modelling who would like to compare alternative models for the same dataset. We illustrate how to plot a graph and the graph obtained by removing an edge using the same layout. To do this, we use the `agopen()` function to generate an `Ragraph` object, which is a representation of the layout of a graph (rather than of the graph as a mathematical object). From this we remove the required edge.

```
> edgeNames(ug3)

[1] "a~b" "a~c" "a~e" "b~c" "c~d" "c~e" "d~e"
> ng3 <- agopen(ug3, name="ug3", layoutType="neato")
> ng4 <- ng3
> AgEdge(ng4) <- AgEdge(ng4)[-3]
> plot(ng3)
```

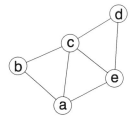

```
> plot(ng4)
```

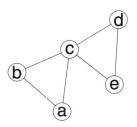

The following example illustrates how individual edge and node attributes may be set. We use the chain graph cG1 described above.

```
> cG1a <- as(cG1, "graphNEL")
> nodes(cG1a) <- c("alpha","theta","tau","beta","pi","upsilon","gamma",
+                   "iota","phi","delta","kappa")
> edges <- buildEdgeList(cG1a)
> for (i in 1:length(edges)) {
+     if (edges[[i]]@attrs$dir=="both") edges[[i]]@attrs$dir <- "none"
+     edges[[i]]@attrs$color <- "blue"
+ }
> nodes <- buildNodeList(cG1a)
> for (i in 1:length(nodes)) {
+     nodes[[i]]@attrs$fontcolor <- "red"
+     nodes[[i]]@attrs$shape <- "ellipse"
+     nodes[[i]]@attrs$fillcolor <- "lightgrey"
+     if (i <= 4) {
+         nodes[[i]]@attrs$fillcolor <- "lightblue"
+         nodes[[i]]@attrs$shape <- "box"
+     }
+ }
> cG1al <- agopen(cG1a, edges=edges, nodes=nodes, name="cG1a",
+                 layoutType="neato")
> plot(cG1al)
```

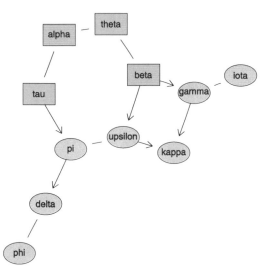

1.4.3 The igraph Package

The **igraph** package is a supplement or alternative to the **graph** package with many neat features. As we have seen, it is easy to convert between graphNEL objects, `igraph` objects and adjacency matrices using the `as()` function. Alternatively `igraph` objects can be created using the `graph.formula()` function:

```
> ug4 <- graph.formula(a -- b:c, c--b:d, e -- a:d)
> ug4

Vertices: 5
Edges: 6
Directed: FALSE
Edges:

[0] 'a' -- 'b'
[1] 'a' -- 'c'
[2] 'b' -- 'c'
[3] 'c' -- 'd'
[4] 'a' -- 'e'
[5] 'd' -- 'e'
> plot(ug4, layout=layout.graphopt)
```

The same graph may be created from scratch as follows:

```
> ug4.2 <- graph.empty(n=5, directed=FALSE)
> V(ug4.2)$name <- V(ug4.2)$label <- letters[1:5]
> ug4.2 <- add.edges(ug4.2, c(0,1, 0,2, 0,4, 1,2, 2,3, 3,4))
> ug4.2

Vertices: 5
Edges: 6
Directed: FALSE
Edges:

[0] 'a' -- 'b'
[1] 'a' -- 'c'
[2] 'a' -- 'e'
[3] 'b' -- 'c'
[4] 'c' -- 'd'
[5] 'd' -- 'e'
```

The graph is displayed using the `plot()` function, with a layout determined using the `graphopt` method. A variety of layout algorithms are available: type `?layout` for an overview. Note that per default the nodes are labelled $0, 1, \ldots$ and so forth. We show how to modify this shortly.

As mentioned previously we have created a custom function `myiplot()` which creates somewhat more readable plots:

```
> myiplot(ug4, layout=layout.graphopt)
```

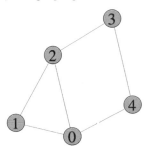

As with `graphNEL` objects, in **igraph** graphs are defined in terms of node and edge lists. In addition, they have *attributes*: these belong to the vertices, the edges or to the graph itself. The following example sets a graph attribute, *layout*, and two vertex attributes, *label* and *color*. These are used when the graph is plotted. The *name* attribute contains the node labels.

```
> ug4$layout    <- layout.graphopt(ug4)
> V(ug4)$label <- V(ug4)$name
> V(ug4)$color <- "red"
> V(ug4)[1]$color <- "green"
> V(ug4)$size <- 40
> V(ug4)$label.cex <- 3
> plot(ug4)
```

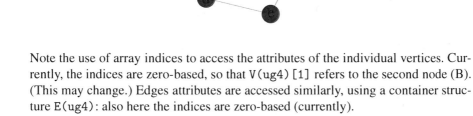

Note the use of array indices to access the attributes of the individual vertices. Currently, the indices are zero-based, so that V(ug4)[1] refers to the second node (B). (This may change.) Edges attributes are accessed similarly, using a container structure E(ug4): also here the indices are zero-based (currently).

It is easy to extend `igraph` objects by defining new attributes. In the following example we define a new vertex attribute, `discrete`, and use this to color the vertices.

```
> ug5 <- set.vertex.attribute(ug4, "discrete", value=c(T,T,F,F,T))
> V(ug5)[discrete]$color <- "green"
> V(ug5)[!discrete]$color <- "red"
> plot(ug5)
```

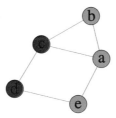

A useful interactive drawing facility is provided with the `tkplot()` function. This causes a pop-up window to appear in which the graph can be manually edited. One use of this is to edit the layout of the graph: the new coordinates can be extracted and re-used by the `plot()` function. For example

```
> tkplot(ug4)
2
```

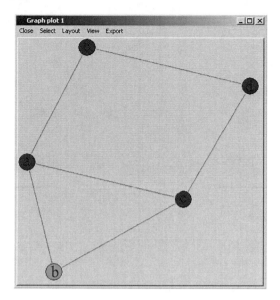

The `tkplot()` function returns a window id (here 2). While the popup window is open, the current layout can be obtained by passing the window id to the `tk-plot.getcoords()` function, as for example

```
> xy <- tkplot.getcoords(2)
> plot(g, layout=xy)
```

It is straightforward to reuse layout information with `igraph` objects. The layout functions when applied to graphs return a matrix of (x, y) coordinates:

```
> layout.spring(ug4)
```

```
           x        y
1  7.568e-01  0.2459
2 -7.568e-01  0.2459
3  4.677e-01 -0.6438
4 -4.677e-01 -0.6438
5  4.337e-19  0.7958
```

Most layout algorithms use a random generator to choose an initial configuration. Hence if we set the layout attribute to be a layout function, repeated calls to plot will use different layouts. For example, after

```
> ug4$layout <- layout.spring
```

repeated invocations of `plot(ug4)` will use different layouts. In contrast, after

```
> ug4$layout <- layout.spring(ug4)
```

the layout will be fixed. The following code fragment illustrates how two graphs with the same vertex set may be plotted using the same layout.

```
> ug5 <- ug(~A*B*C + B*C*D + D*E, result='igraph')
> ug6 <- ug(~A*B+B*C+C*D+D*E, result='igraph')
> ug6$layout         <- ug5$layout         <- layout.spring(ug5)
```

```
> V(ug5)$size        <- V(ug6)$size        <- 50
> V(ug5)$label.cex <- V(ug6)$label.cex <- 3
> par(mfrow=c(1,2))
> plot(ug5); plot(ug6)
```

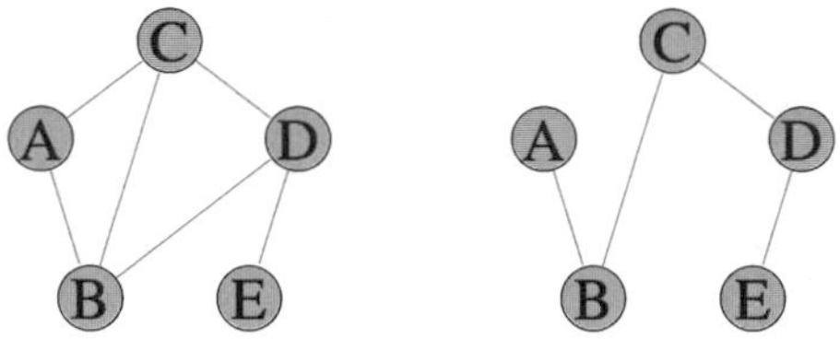

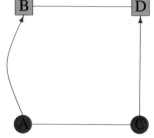

An overview of attributes used in plotting can be obtained by typing
?igraph.plotting. A final example illustrates how more complex graphs can
be displayed:

```
> em1 <- c(0,0,1,0,1,0,0,1,1,0,0,0,1,1,1,0)
> dim(em1) <- c(4,4)
> iG  <- graph.adjacency(em1)
> V(iG)$shape <- c("circle","square","circle","square")
> V(iG)$color <- c("red","green")
> V(iG)$label <- c("A", "B", "C", "D")
> E(iG)$arrow.mode <- c(2,0)[1+is.mutual(iG)]
> E(iG)$color  <- c("blue", "black")
> E(iG)$curved <- c(T,F,F,F,F,F)
> iG$layout    <- layout.graphopt(iG)
> myiplot(iG)
```

An image showing a graph with nodes B, D (squares) at top and A, C (circles) at bottom.

1.4.4 3-D Graphs

The gplot3d() function in the **sna** package displays a graph in three dimensions.
Using a mouse, the graph can be rotated and zoomed. Opinions differ as to how
useful this is. The following code fragment can be used to try the facility. First
we derive the adjacency matrix of a built-in graph in the **igraph** package, then we
display it as a (pseudo)-three-dimensional graph.

```
> library(sna)
> aG <- as(graph.famous("Meredith"),"matrix")
> gplot3d(aG)
```

1.4.5 Alternative Graph Representations

As mentioned above, graphNEL objects are so-called because they use a node and edge list representation. So these can also be created directly, by specifying a vector of nodes and a list containing the edges corresponding to each node. For example,

```
> V <- c("a","b","c","d")
> edL <- vector("list", length=4)
> names(edL) <- V
> for (i in 1:4) {
+    edL[[i]] <- list(edges=5-i)
+ }
> gR <- new("graphNEL", nodes=V, edgeL=edL)
> plot(gR)
```

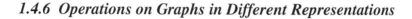

1.4.6 Operations on Graphs in Different Representations

The functions for operations on graphs illustrated in the previous sections are all available for graphs in the graphNEL representation (some operations are in fact available for graphs in the other representations as well). Notice that the functions differ in whether they take the graph as the first or as the last argument (that is mainly related to different styles in different packages).

The **gRbase** package has a function querygraph() which provides a common interface to the graph operations for undirected graphs and DAGs illustrated above. Moreover, querygraph() works on graphs represented as graphNEL objects, igraph objects and adjacency matrices. The general syntax is

```
> args(querygraph)
```
```
function (object, op, set = NULL, set2 = NULL, set3 = NULL)
NULL
```

For example, we obtain:

```
> ug_NEL <- ug(~a:b+b:c:d+e)
> ug_igraph <- as(ug_NEL, "igraph")
> separates("a","d",c("b","c"), ug_NEL)
```
```
[1] TRUE
```
```
> querygraph(ug_igraph, 'separates', "a","d",c("b","c"))
```
```
[1] TRUE
```

Chapter 2
Log-Linear Models

2.1 Introduction

This chapter gives an account of graphical models for multivariate discrete data. Such data are usually summarized as contingency tables, and Sect. 2.2 describes some general utilities useful when working with such tables. Section 2.3 introduces the theory of log-linear models, illustrating this using dModel objects from the **gRim** package. Section 2.5.1 shows how log-linear models can be fit using the glm function, and Sect. 2.5.2 describes some aspects of working with dModel objects. Some more advanced topics are dealt with in Sect. 2.5.

2.2 Preliminaries

2.2.1 Four Datasets

To introduce contingency table data we consider four examples. All datasets used here are in **gRbase**. The first is shown in Table 2.1. These data originate from Schoener (1968) and are discussed in numerous places, e.g. Edwards (2000) and Whittaker (1990). In a study of lizard behaviour, characteristics of 409 lizards were recorded, namely species (S), perch diameter (D) and perch height (H). The focus of interest is in how the propensities of the lizards to choose perch height and diameter are related, and whether and how these depend on species.

The second dataset we consider is a 2^6 contingency table concerning risk factors for coronary heart disease. The data originated in a prospective study of coronary heart disease carried out in Czechoslovakia (Reiniš et al. 1981). For a sample of 1841 car-workers, the following information was recorded: whether they smoked, whether their work was strenuous mentally, whether their work was strenuous physically, whether their systolic blood pressure was less than 140 mm), whether the ratio of beta to alpha lipoproteins was less than 3, and whether there was a family history of coronary heart disease.

S. Højsgaard et al., *Graphical Models with R*, Use R!,
DOI 10.1007/978-1-4614-2299-0_2, © Springer Science+Business Media, LLC 2012

Table 2.1 Perching behaviour of two species of lizards

Species	Perch diameter (inches)	Perch Height (feet)	
		> 4.75	≤ 4.75
Anoli	≤ 4	32	86
	> 4	11	35
Distichus	≤ 4	61	73
	> 4	41	70

```
> data(reinis)
> str(reinis)

 table [1:2, 1:2, 1:2, 1:2, 1:2, 1:2] 44 40 112 67 129 145 12 23 35 12 ...
 - attr(*, "dimnames")=List of 6
  ..$ smoke  : chr [1:2] "y" "n"
  ..$ mental : chr [1:2] "y" "n"
  ..$ phys   : chr [1:2] "y" "n"
  ..$ systol : chr [1:2] "y" "n"
  ..$ protein: chr [1:2] "y" "n"
  ..$ family : chr [1:2] "y" "n"
```

The third dataset is a 2^6 contingency table taken from genetics, and analyzed in Edwards (2000). Two isolates of the barley powdery mildew fungus were crossed, and for 70 progeny 6 binary characteristics (genetic markers) were recorded.

```
> data(mildew)
> str(mildew)

 table [1:2, 1:2, 1:2, 1:2, 1:2, 1:2] 0 0 0 0 3 0 1 0 0 1 ...
 - attr(*, "dimnames")=List of 6
  ..$ la10: chr [1:2] "1" "2"
  ..$ locc: chr [1:2] "1" "2"
  ..$ mp58: chr [1:2] "1" "2"
  ..$ c365: chr [1:2] "1" "2"
  ..$ p53a: chr [1:2] "1" "2"
  ..$ a367: chr [1:2] "1" "2"
```

The fourth dataset is a three-way table containing the results of a study comparing four different surgical operations on patients with duodenal ulcer, carried out in four centres, and described in Grizzle et al. (1969). The four operations were: vagotomy and drainage, vagotomy and antrectomy (removal of 25% of gastric tissue), vagotomy and hemigastrectomy (removal of 50% of gastric tissue), and gastric restriction (removal of 75% of gastric tissue). The response variable is the severity of gastric dumping, an undesirable syndrome associated with gastric surgery.

```
> data(dumping)
> str(dumping)

 table [1:3, 1:4, 1:4] 23 7 2 23 10 5 20 13 5 24 ...
 - attr(*, "dimnames")=List of 3
  ..$ Symptom  : chr [1:3] "none" "slight" "moderate"
  ..$ Operation: chr [1:4] "Vd" "Va" "Vh" "Gr"
  ..$ Centre   : chr [1:4] "1" "2" "3" "4"
```

The first and second variables are ordinal.

2.2.2 Data Formats

Multivariate discrete data are usually stored in one of three forms, here illustrated with the `lizard` data.

As a Raw Case-List   For example, the lizard data could be represented as 409 observations of three discrete variables: species, perch diameter and perch height. This is typically represented in R as a dataframe, with the discrete variables represented as factors. For example,

```
> data(lizardRAW)
> head(lizardRAW)

  diam height species
1   >4  >4.75    dist
2   >4  >4.75    dist
3  <=4 <=4.75   anoli
4   >4 <=4.75   anoli
5   >4 <=4.75    dist
6  <=4 <=4.75   anoli
```

As an Aggregated Case-List   Sometimes discrete data are represented in aggregated *case-list* form (again typically represented as a `data.frame` in R), where one variable (usually called `Freq`) stores the counts for each configuration of variables:

```
> data(lizardAGG)
> lizardAGG

  diam height species Freq
1  <=4  >4.75   anoli   32
2   >4  >4.75   anoli   11
3  <=4 <=4.75   anoli   86
4   >4 <=4.75   anoli   35
5  <=4  >4.75    dist   61
6   >4  >4.75    dist   41
7  <=4 <=4.75    dist   73
8   >4 <=4.75    dist   70
```

As a Contingency Table   Another aggregated representation of data is as a *contingency table* (which in R is represented as a `table` or as an `array`):

```
> data(lizard)
> lizard

, , species = anoli

     height
diam  >4.75 <=4.75
  <=4    32     86
  >4     11     35

, , species = dist

     height
```

```
diam  >4.75  <=4.75
  <=4    61      73
  >4     41      70
```

Note that the contingency table form is a compact representation of data when these are dense, in the sense that the number of observations is larger than the number of combinations of variable levels. With sparse data, for which the number of combinations of variable levels exceeds the number of observations, the case list format is more compact.

Note that coercion between the different representations can be obtained as follows:

```
> ##
> ## Raw case-list to aggregated case-list:
> as.table(ftable(lizardRAW))
> ##
> ## Raw case-list to table
> xtabs(~., data=lizardRAW)
> ##
> ## Aggregated case-list to table
> xtabs(Freq~., data=lizardAGG)
> ##
> ## Table to aggregated case--list
> as.data.frame(lizard)
```

Note also that the lizard data can be specified as a contingency table using

```
> counts <- c(32, 11, 86, 35, 61, 41, 73, 70)
> dimn <- list(diam=c("<=4", ">4"),
+              height=c(">4.75", "<=4.75"),
+              species=c("anoli", "dist"))
> lizard <- as.table(array(counts, dim=c(2,2,2), dimnames=dimn))
```

2.3 Log-Linear Models

In this section we give a brief account of the theory of log-linear models.

2.3.1 Preliminaries and Notation

Suppose that we have a dataset with N observations of d discrete random variables. For example, the lizard data had $N = 409$ and $d = 3$. We write the collection of discrete variables as $X = (X_v)_{v \in \Delta}$, and we call the possible values a discrete variable may take its *levels*. Write the number of levels of X_v as $|X_v|$. For notational convenience we label the levels $1, \ldots, |X_v|$ though in practice they should be given more meaningful labels. We can then write a generic observation (or *cell*) as $i = (i_1, \ldots, i_d)$, and the set of possible cells as $\mathcal{I}$.

We assume that the observations are independent and are interested in modelling the probabilities $p(i) = \Pr(X = i)$ for $i \in \mathcal{I}$. The joint probability of the observations represented as a case list $(i^\nu, \ \nu = 1, \ldots, N)$ is then

$$p(i^\nu, \nu = 1, \ldots, N) = \prod_{\nu=1}^{N} p(i^\nu) = \prod_{i \in \mathcal{I}} p(i)^{n(i)} \tag{2.1}$$

where we have formed an aggregated case list or, equivalently, the contingency table $\{n(i)\}_{i \in \mathcal{I}}$, where $n(i)$ is the number of cases i^ν with $i^\nu = i$. The joint probability of the observed contingency table is

$$p(\{n(i)\}_{i \in \mathcal{I}}) = \frac{N!}{\prod_{i \in \mathcal{I}} n(i)!} \prod_{i \in \mathcal{I}} p(i)^{n(i)} \tag{2.2}$$

which differs from (2.1) by a multinomial coefficient which does not affect the likelihood as the latter is only determined up to a constant factor.

$$L(p) \propto \prod_{i \in \mathcal{I}} p(i)^{n(i)}. \tag{2.3}$$

If we do not restrict the probabilities in any way (except requiring that they are non-negative and sum to unity), then it is easily shown that the maximum likelihood estimates are given by $\hat{p}(i) = n(i)/N$ for $i \in \mathcal{I}$. The unrestricted model is known as the *saturated model*. In most substantive contexts it is of interest to restrict the probabilities further to obtain parsimony and to identify or exploit structural information, see further in Sect. 2.3.2 below.

We need a little more notation. The expected cell counts are written $m(i) = Np(i)$ for $i \in \mathcal{I}$, and the fitted values as $\hat{m}(i) = N\hat{p}(i)$. We need to work with marginal tables and to do this must first define marginal cells. Recall that Δ contains d variables and so a generic cell i is a d-tuple, that is $i = (i_1, \ldots, i_d)$. For a subset $A \subseteq \Delta$, the corresponding marginal cell is written i_A, and contains the indices $(i_\nu, \nu \in A)$. The corresponding marginal counts and probabilities are written $n(i_A)$ and $p(i_A)$. So, for example, we have that $n(i_A) = \sum_{j \in \mathcal{I}: j_A = i_A} n(j)$.

2.3.2 Hierarchical Log-Linear Models

Log-linear models are defined by constraining the logarithms of the probabilities to follow ANOVA-like factorial expansions. For example, for a three-dimensional table (such as Table 2.1) we write a generic cell as $i = (j, k, l)$, the variables as $\Delta = \{a, b, c\}$, and the saturated model as

$$\log p(i) = u + u_j^a + u_k^b + u_l^c + u_{jk}^{ab} + u_{jl}^{ac} + u_{kl}^{bc} + u_{jkl}^{abc}. \tag{2.4}$$

Here the u's are unknown parameters—usually called *interaction terms*. To estimate these uniquely we would need to constrain them further in some way, but we do not need to bother about this now.

The expansion (2.4) is only valid when $p(i) > 0$. We obtain higher generality by letting $\tilde{u} = \exp u$ and writing the expansion in product form

$$p(i) = \tilde{u} \cdot \tilde{u}_j^a \cdot \tilde{u}_k^b \cdot \tilde{u}_l^c \cdot \tilde{u}_{jk}^{ab} \cdot \tilde{u}_{jl}^{ac} \cdot \tilde{u}_{kl}^{bc} \cdot \tilde{u}_{jkl}^{abc} \tag{2.5}$$

as this enables us to deal with cells i with $p(i) = 0$. In general we have to includes limits of distributions satisfying (2.4) or (2.5), see further discussion in Sects. 2.3.4 and 2.5.1 below.

In log-linear models, certain interaction terms are set to zero. For example, we could set all two- and three-factor interaction terms equal to zero, by positing that

$$\log p(i) = u + u_j^a + u_k^b + u_l^c.$$

This is called the *main effect* model.

Usually only hierarchical log-linear models are of interest. The term *hierarchical* means that if a term is set to zero, all its higher-order relatives are also set to zero. Alternatively expressed, if a term is allowed in the expansion, so are other terms of lower order involving the relevant variables. For example, if we set $u_{jk}^{ab} = 0$ for all j, k, then we must also set $u_{jkl}^{abc} = 0$ for all j, k, l and if $u_{jk}^{ab} \neq 0$ is allowed, we must allow $u_j^a \neq 0$ and $u_k^b \neq 0$.

Hierarchical models can be specified in terms of the maximal interaction terms permitted: these are called the *generators* of the model. For example, the generators of the model

$$\log p(i) = u + u_j^a + u_k^b + u_l^c + u_{jk}^{ab} + u_{jl}^{ac} \tag{2.6}$$

are $\{a, b\}$ and $\{a, c\}$.

The **gRim** package has a function dmod() to define and fit hierarchical log-linear models. The models can be specified using a model formula or list of character vectors representing the generators. For example,

```
> m1 <- dmod(~species*height+species*diam, data=lizard)
> m2 <- dmod(list(c("species","height"),c("species", "diam")),
             data=lizard)
```

specify the same model. The first form is most useful when specifying small models by hand.

Under (2.6) specified as m1 or m2 above the probabilities can be factored into

$$p(i) = (\tilde{u} \cdot \tilde{u}_j^a \cdot \tilde{u}_k^b \cdot \tilde{u}_{jk}^{ab})(\tilde{u}_l^c \cdot \tilde{u}_{jl}^{ac}),$$

i.e. into two factors, the first not involving c and the second not involving b. It then follows from the factorization criterion (1.1) that $b \perp\!\!\!\perp c \mid a$. More generally this reasoning implies that under any hierarchical model, two factors are conditionally independent given the rest if and only if the corresponding two-factor interaction term is set to zero or, equivalently, if no generator contains both factors.

Thus, this model implies that perching diameter and height are independent given species. In other words, for each species considered separately, perching diameter and height are independent.

The *dependence graph* of a hierarchical model is an undirected graph with edges present whenever the corresponding two-factor interaction is allowed. We can display the graph of a dModel object using plot (see Fig. 2.1).

Fig. 2.1 Conditional
independence of `diam` and
`height` given `species`

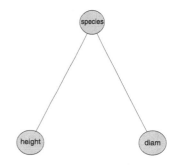

From the global Markov property (Sect. 1.3) we can find out which conditional independences hold under a model:

```
> separates("height","diam","species", as(m1,"graphNEL"))
```

```
[1] TRUE
```

In the present case the property is evident from the graph, but the facility is useful for higher-dimensional models.

2.3.3 Graphical and Decomposable Log-Linear Models

Suppose that we are given an undirected graph $G = (\Delta, E)$, and consider the hierarchical log-linear model $\mathcal{M}$ for Δ whose generators are identical to the cliques of the graph. A model that can be specified in this way is called a *graphical* model. Since the two-factor interaction terms that are set to zero in the model correspond to edges that are not present in G, we see that G is the dependence graph of $\mathcal{M}$. By the hierarchical principle, any higher-order interaction term containing such a 'zero' two-factor term is also set to zero. And any higher-order term that does not contain a 'zero' two-factor term is contained in a generator and so is not set to zero. So one characterization of a graphical log-linear model is that the two-factor interaction terms present in the model completely determine which higher-order interactions are also present. Log-linear models that are not graphical set higher order interactions to zero even though all the corresponding two-factor interactions are not set to zero. The simplest *non-graphical* model is the no three-factor interaction model for a three-way table:

```
> no3f <- dmod(~species:height + species:diam + height:diam,
               data=lizard)
```

Note that this model has the same dependence graph as the saturated model:

```
> par(mfcol=c(1,2))
> sat <- dmod(~species:height:diam, data=lizard)
> plot(no3f, main='no 3-factor interaction')
> plot(sat, main='saturated model')
```

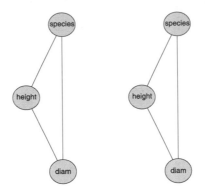

The attractive feature of graphical models is that they can be interpreted solely in terms of patterns of conditional independences, which can be displayed in terms of a graph.

We can also obtain the graphical model corresponding to a given undirected graph, as in

```
> g <- ug(~la10:locc:mp58 + locc:mp58:c365 + mp58:c365:p53a +
+          c365:p53a:a367)
> mg <- dmod(g, data=mildew)
```

In general, to obtain maximum likelihood estimates for log-linear models, iterative methods must be used. But for an important subclass of log-linear models, the *decomposable* models, closed-form expressions are available; see Sect. 2.3.4 for details.

To get information about properties of a model, the summary() method may be used:

```
> summary(no3f)
```

```
is graphical=FALSE; is decomposable=FALSE
generators (glist):
  :"species" "height"
  :"species" "diam"
  :"height" "diam"
```

2.3.4 Estimation, Likelihood, and Model Fitting

Decomposable models are characterized as graphical models whose graphs are triangulated. For decomposable models, closed-form expressions exist for the maximum likelihood estimate. The closed-form expressions are closely related to RIP-orderings of the cliques; see Sect. 1.4.1 for further details.

Let $\mathcal{C} = (C_1, \ldots, C_k)$ be such an ordering and $\mathcal{S} = (S_1, \ldots, S_k)$ the corresponding separators. Then the ML estimator is given by

$$\hat{m}(i) = \frac{\prod_{j=1\ldots k} n(i_{C_j})}{\prod_{j=1\ldots k} n(i_{S_j})}.$$

For non-decomposable models we need another way to find the maximum likelihood estimates. Most commonly the IPS (iterative proportional scaling) algorithm is used. This is a simple and robust algorithm, which works by storing and iteratively updating a table of fitted values $\{m(i)\}_{i \in \mathcal{I}}$.

Let $\mathcal{C} = \{a_1, \ldots, a_Q\}$ be the generators of a hierarchical log-linear model. The corresponding marginal tables $n(i_{a_k})$, $k = 1, \ldots, Q$, are a set of sufficient statistics. The maximum likelihood estimate is obtained by equating the sufficient statistics with their expectations $m(i_{a_k})$.

Initially the $m(i)$ are set to some constant, say $m(i) = 1$ for all $i \in \mathcal{I}$. One iteration consists of updating for each $k = 1, \ldots, Q$

$$m(i) \leftarrow m(i) \frac{n(i_{a_k})}{m(i_{a_k})} \quad \forall i \in \mathcal{I}. \tag{2.7}$$

Iteration continues until convergence which happens when $m(i_{a_k}) = n(i_{a_k})$. The algorithm is always theoretically convergent with the limiting value being the maximum likelihood estimate under the model $\{\hat{m}(i)\}_{i \in \mathcal{I}}$, although these may not all have $\hat{m}(i) > 0$ for all cells i and thus may not admit a logarithmic expansion.

In R, the IPS algorithm is implemented in the `loglin()` function: the function `dmod()` in the **gRim** package provides an interface to this.

Notice that if the cliques of a decomposable model are given such that they follow a RIP-ordering then the IPS algorithm will converge after one iteration. If the cliques do not follow a RIP-ordering then IPS will converge after two iterations.

A disadvantage of IPS is that for high-dimensional problems it can be computationally expensive to store and update the whole table, as the iteration (2.7) passes through all possible values of i. It is possible to avoid this using message passing techniques based on the factorization (2.5), similar to those implemented in **gRain** and described in Chap. 3.

Another algorithm is that of iteratively reweighted least squares which is used for generalized linear models. This alternative is attractive when there is interest in the log-linear parameters (u-terms) themselves, since as a byproduct it provides estimates and standard errors of these. However, this approach can be problematic for other reasons; see Sect. 2.5.1 for an example and further discussion.

2.3.5 Hypothesis Testing

The maximized log-likelihood of a model m is given, up to an arbitrary additive constant, by

$$\ell = \sum_{i \in \mathcal{T}} n(i) \log \hat{p}(i)$$

where $\hat{p}(i)$ are the maximum likelihood estimates.

The *deviance* of a model $\mathcal{M}$ is twice the log-likelihood ratio of $\mathcal{M}$ versus the saturated model, i.e.,

$$D = \text{dev} = 2(\hat{\ell}_s - \hat{\ell}_m),$$

where $\hat{\ell}_s$ and $\hat{\ell}_m$ are the maximized log-likelihoods under the saturated model and $\mathcal{M}$, respectively. In this case we obtain

$$D = \text{dev} = G^2 = 2\sum_{i \in \mathcal{I}} n(i) \log \frac{n(i)}{\hat{m}(i)}.$$

Under $\mathcal{M}$, D is asymptotically $\chi^2(k)$ where the degrees of freedom k is the difference in dimension (number of free parameters) between the saturated model and m. So the deviance provides a goodness-of-fit test for the model. For example the following model fits rather well:

```
> m1 <- dmod(~species:height+species:diam, data=lizard)
> m1

Model: A dModel with 3 variables
 graphical :   TRUE  decomposable :   TRUE
 -2logL    :        1604.43 mdim :    5 aic :      1614.43
 ideviance :          23.01 idf  :    2 bic :      1634.49
 deviance  :           2.03 df   :    2
```

An alternative to the deviance is Pearson's goodness-of-fit test, defined by

$$X^2 = \sum_{i \in \mathcal{I}} \frac{\{n(i) - \hat{m}(i)\}^2}{\hat{m}(i)}$$

which has the same asymptotic distribution under the null hypothesis. This can be obtained using

```
> m1$fitinfo$pearson
```

```
[1] 2.017
```

Notice that it follows from the general definition of deviance given above that to calculate the deviance, it must be possible to fit the saturated model. This can always be done for log-linear models, but may not be possible in general, for example for Gaussian models; see Chap. 4. When working with sparse graphical models it is therefore often simpler to consider the *ideviance* (or independence deviance) which we define as twice the log-likelihood ratio between the model in question and the model of complete independence, corresponding to a graph with no edges, i.e. in this case

$$iD = \text{idev} = 2\sum_{i \in \mathcal{I}} n(i) \log \frac{\hat{m}(i)}{\prod_{v \in V} n(i_v)}.$$

The deviance or ideviance *difference* between two nested models makes always sense, provided both can be fitted, and it is in both cases equal to twice the log-likelihood ratio.

A related issue is that the dimension of a model depends, strictly speaking, on the sampling scheme employed, whereas the difference in dimension between two nested models (i.e. the degrees of freedom) does not. As we have described it, data have been assumed to be collected as a fixed number of independent units, referred to as the *multinomial sampling scheme*. If we instead assume that the total number of observations follows a Poisson distribution with unknown parameter $\lambda > 0$, the counts $N(i)$ become independent with parameters $e\{N(i)\} = m(i)$. This is the *Poisson sampling scheme*.

It can be shown that the maximum likelihood estimate of λ is then equal to $\hat{\lambda} = n$ and the likelihood function for $\lambda = \hat{\lambda}$ is proportional to the likelihood function in the multinomial sampling scheme, thus not affecting deviances nor maximum likelihood estimates. This is known as the *Poisson trick*.

In general, it is simplest to calculate dimensions of models for the Poisson sampling scheme and therefore all dimensions refer to this scheme.

The calculation of the model dimension by dmod() assumes that $\hat{m}(i) > 0$ for all cells i, which will be the case when the data are dense, for example when all the cell counts are positive. When the data are sparse, as is usually the case for moderate to high-dimensional problems, some cells may have $\hat{m}(i) = 0$ and so the degrees of freedom shown need adjustment. Calculation of the appropriate degrees of freedom when the data are sparse is a hard problem, and we are aware of no software that does this correctly in all cases. In any case the asymptotic χ^2 approximation may be poor if $\hat{m}(i)$ is small.

A viable approach to analysis is to focus on comparisons of nested decomposable models, for which the correct adjustment to the degrees of freedom can be calculated, and for which it is straightforward to calculate exact conditional tests. A key result is that if $\mathcal{M}_0 \subset \mathcal{M}_1$ are decomposable log-linear models differing by one edge $e = \{u, v\}$ only, then e is contained in one clique C of $\mathcal{M}_1$ only, and the likelihood ratio test for $\mathcal{M}_0$ versus $\mathcal{M}_1$ can be performed in the marginal C-table as a test of $u \perp\!\!\!\perp v \mid C \setminus \{u, v\}$. The point being partly that the marginal table on C may not be sparse, but more importantly, that it is straightforward to adjust the degrees of freedom for a pure test of conditional independence like this, as we describe shortly.

For example, suppose that we specify a decomposable model m3 for the mildew data, and delete the edge {locc, a367} from m3 using the update() function, obtaining a model m4. This function is described below in Sect. 2.5.2. The edge deleted is contained in one clique only of m3, so m4 is also decomposable.

```
> m3 <- dmod(~la10*locc*mp58*c365*p53a+locc*mp58*c365*p53a*a367,
            data=mildew)
> m4 <- update(m3, list(dedge=~locc*a367))
> oldpar<-par(mfrow=c(1,2))
> plot(m3, "neato")
> plot(m4, "neato")
> par(oldpar)
```

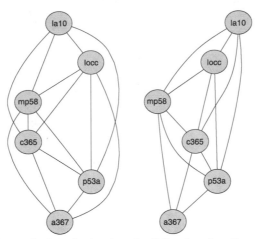

A direct comparison of m3 and m4 using the following function gives an incorrect value for the degrees of freedom

```
>    comparemodels <- function(m1,m2) {
+    lrt <- m2$fitinfo$dev - m1$fitinfo$dev
+    dfdiff <- m1$fitinfo$dimension[1] - m2$fitinfo$dimension[1]
+    c('lrt'=lrt, 'df'=dfdiff)
+ }
> m3
```

```
Model: A dModel with 6 variables
 graphical :  TRUE  decomposable :  TRUE
 -2logL    :         366.16 mdim :   47 aic :       460.16
 ideviance :         209.32 idf  :   41 bic :       565.83
 deviance  :           0.40 df   :   16
Notice: Table is sparse
  Asymptotic chi2 distribution may be questionable.
  Degrees of freedom can not be trusted.
  Model dimension adjusted for sparsity : 21
```

```
> m4
```

```
Model: A dModel with 6 variables
 graphical :  TRUE  decomposable :  TRUE
 -2logL    :         370.73 mdim :   39 aic :       448.73
 ideviance :         204.74 idf  :   33 bic :       536.42
 deviance  :           4.98 df   :   24
Notice: Table is sparse
  Asymptotic chi2 distribution may be questionable.
  Degrees of freedom can not be trusted.
  Model dimension adjusted for sparsity : 22
```

```
> comparemodels(m3,m4)
```

```
      lrt df.mod.dim
    4.573      8.000
```

The correct test may be obtained using the testdelete() function:

```
> testdelete(m3, edge=c("locc","a367"))
```

```
dev:    4.573 df:   3 p.value: 0.20585 AIC(k=2.0):    -1.4 edge: locc:a367
host:   locc mp58 c365 p53a a367
Notice: Test performed in saturated marginal model
```

This function identifies that m3 is decomposable and that the edge {locc, a367} is in one clique C only. The test is then performed as a test of $u \perp\!\!\!\perp v \mid C \setminus \{u, v\}$. Note that the test statistic matches with that of comparemodels() and the degrees of freedom have been correctly adjusted.

Tests of general conditional independence hypotheses of the form $u \perp\!\!\!\perp v \mid W$ can be performed using the ciTest_table() function.

```
> cit <- ciTest_table(mildew, set=c("locc","a367","mp58","c365",
                                    "p53a"))

Testing locc _|_ a367 | mp58 c365 p53a
Statistic (DEV):    4.573 df: 3 p-value: 0.2059 method: CHISQ
```

The general syntax of the set argument is of the form (u, v, W) where u and v are variables and W is a set of variables. The set argument can also be given as a right-hand sided formula.

Notice that in this case the results are identical to those given by the test-delete() function, since we have specified the correct conditioning set. If we had conditioned on more variables

```
> cit2 <- ciTest_table(mildew, set=c("locc","a367","mp58","c365",
+                                    "p53a","la10"))

Testing locc _|_ a367 | mp58 c365 p53a la10
Statistic (DEV):    4.553 df: 3 p-value: 0.2076 method: CHISQ
```

different results would be obtained.

In model terms, the test performed by ciTest_table() corresponds to the test for removing the edge $\{u, v\}$ from the saturated model with variables $\{u, v\} \cup W$. If we (conceptually) form a factor S by crossing the factors in W, we see that the test can be formulated as a test of the conditional independence $u \perp\!\!\!\perp v \mid S$ in a three way table. The deviance decomposes into independent contributions from each stratum:

$$
D = 2 \sum_{ijs} n_{ijs} \log \frac{n_{ijs}}{\hat{m}_{ijs}}
$$

$$
= \sum_s 2 \sum_{ij} n_{ijs} \log \frac{n_{ijs}}{\hat{m}_{ijs}} = \sum_s D_s
$$

where the contribution D_s from the sth slice is the deviance for the independence model of u and v in that slice. For example,

```
> cit$slice
```

```
  statistic p.value df mp58 c365 p53a
1    0.0000 1.00000  0    1    1    1
2    0.5053 0.47716  1    2    1    1
3    1.2953 0.25508  1    1    2    1
4    2.7726 0.09589  1    2    2    1
5    0.0000 1.00000  0    1    1    2
6    0.0000 1.00000  0    2    1    2
```

```
7      0.0000 1.00000  0    1    2    2
8      0.0000 1.00000  0    2    2    2
```

The sth slice is a $|u| \times |v|$ table $\{n_{ijs}\}_{i=1\ldots|u|, j=1\ldots|v|}$. The output shows the degrees of freedom corresponding to the test for independence in each slice, given by

$$df_s = (\#\{i : n_{i \cdot s} > 0\} - 1)(\#\{j : n_{\cdot js} > 0\} - 1)$$

where $n_{i \cdot s}$ and $n_{\cdot js}$ are the marginal totals. So the correct number of degrees of freedom for the test in the present example is not 8 but 3, as calculated by the ciTest_table() and testdelete() functions.

An alternative to the asymptotic χ^2 test is to determine the reference distribution using Monte Carlo methods. The marginal totals are sufficient statistics under the null hypothesis, and in a conditional test the test statistic is evaluated in the conditional distribution given the sufficient statistics. Hence one can generate all possible tables with those given margins, calculate the desired test statistic for each of these tables and then see how extreme the observed test statistic is relative to those of the calculated tables. A Monte Carlo approximation to this procedure is to randomly generate a large number of tables with the given margins, evaluate the statistic for each simulated table and then see how extreme the observed test statistic is in this distribution. This is called a *Monte Carlo exact test* and it provides a *Monte Carlo p-value*. In the present example we get a Monte Carlo p-value which is considerably larger than the asymptotic one:

```
> ciTest_table(mildew, set=c("locc","a367","mp58","c365","p53a"),
+ method='MC')

Testing locc _|_ a367 | mp58 c365 p53a
Statistic (DEV):    4.573 df: NA p-value: 0.5550 method: MC
```

An advantage of the Monte Carlo method is that any test statistic can be used, so statistics that are sensitive to specific forms of deviation from independence can be used. In particular, when one or both of u and v are ordinal, more powerful tests of $u \perp\!\!\!\perp v \mid S$ can be applied. The ciTest_ordinal() function supports this approach for three rank tests: the Wilcoxon, Kruskal-Wallis and Jonckheere-Terpstra tests. The Wilcoxon test is applicable when u is binary and v ordinal; the Kruskal-Wallis test when u is nominal and v is ordinal; and the Jonckheere-Terpstra test when both u and v are ordinal. We illustrate use of the function using the dumping syndrome data described above in Sect. 2.2.1. Recall that the three variables are Symptom, Operation and Centre. The first two are ordinal and the third is nominal.

```
> ciTest_ordinal(dumping,c(2,1,3),"jt", N=1000)

$JT
[1] 9566

$EJT
[1] 8705
```

```
$P
[1] 0.009804

$montecarlo.P
[1] 0.005

$set
[1] "Operation"  "Symptom"    "Centre"

> ciTest_ordinal(dumping,c(2,1,3),"deviance", N=1000)

$deviance
[1] 23.54

$df
[1] 24

$P
[1] 0.4883

$montecarlo.P
[1] 0.585

$set
[1] "Operation"  "Symptom"    "Centre"
```

The second argument is a vector of column numbers (if a dataframe is supplied) or dimension numbers (if a table is supplied, as here) of $\{u, v, S\}$. The corresponding names may also be given. The function calculates the Monte Carlo p-value based on N random samples, together with the asymptotic p-value. If $N = 0$, only the latter is calculated. We see that the ordinal test strongly rejects the hypothesis that Symptom is independent of Operation given Centre, whereas the non-ordinal test finds no evidence against this. In this example, the Monte Carlo p-values are similar to the asymptotic ones. To examine whether the conditional distribution of Symptom given Operation is homogeneous over the centres, the Kruskal-Wallis test may be used:

```
> ciTest_ordinal(dumping, c(3,1,2),"kruskal", N=1000)

$KW
[1] 10.02

$df
[1] 12

$P
[1] 0.6143

$montecarlo.P
[1] 0.615

$set
[1] "Centre"    "Symptom"    "Operation"
```

The distributions appear to be homogeneous.

2.4 Model Selection

Using graphs to represent models has the effect of shifting the emphasis from estimation of parameters for a given model towards estimation of the model structure, that is, selecting an appropriate model. Model selection is challenging, not least because the number of possible models is huge. For example, the number of undirected graphs with 30 nodes is $2^{30 \times 29/2} = 2^{435} > 10^{80}$, the estimated number of atoms in the observable universe.

Many different methods to select graphical models have been proposed, but generally they fall into three categories:

- Use of low-order conditional independence tests to infer the structure of the joint model. An example is the PC algorithm (Sect. 4.6.1).
- Heuristic search to optimize some criterion. Often local search around a current model is used to find a local optimum, possibly with combined with a stochastic search method. An example is the hill-climbing algorithm described in Sect. 4.6.2.1.
- Bayesian methods, often involving Markov chain Monte Carlo methods. We do not discuss Bayesian approaches to model selection further, but in Chap. 6 we describe aspects of graphical models from a Bayesian perspective.

Sometimes the first type of methods are used in a preliminary phase and then combined with others for refinement.

The **gRim** package implements a popular variant of the second type using well-known model selection criteria of *penalized likelihood* type. Consider a set of models $\mathcal{M}(j)$ for $j = 0, 1, \ldots, R$. We select the model $\mathcal{M}(j)$ which minimizes $-2 \log L(j) + k p(j)$, where $p(j)$ is the number of free parameters in model $\mathcal{M}(j)$ and k is a penalty parameter.

Akaike's Information Criterion or *AIC* (Akaike 1974) uses $k = 2$. A popular alternative is the *Bayesian Information Criterion* or *BIC* (Schwarz 1978), which sets k to the logarithm of the number of observations. Use of a larger k penalizes complex models more heavily, and so tends to select simpler models. Other values of k can be chosen. It is standard usage in R to call the criterion AIC, even though strictly speaking only the value $k = 2$ gives the "genuine AIC".

The `stepwise()` function searches by default incrementally from an initial model, adding or deleting the edge that gives the largest decrease in the AIC. If there is none the process stops. The search is directional: either forward (adding edges) or backward (deleting edges). Alternatively, significance tests can be used to judge the relative adequacy of the models compared.

The following code selects a model for the `reinis` dataset. The initial model is set to be the saturated model, using a model specification shortcut described in Sect. 2.5.2.

```
> m.init <- dmod(~.^., data=reinis)
> m.reinis <- stepwise(m.init)
> plot(m.reinis)
```

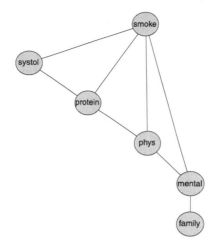

The penalty term k is by default 2, but this can be changed using the argument of the same name. For example, the BIC criterion uses the logarithm of the number of observations as the penalty term:

```
> m.reinis.2  <- stepwise(m.init,k=log(sum(reinis)))
> plot(m.reinis.2)
```

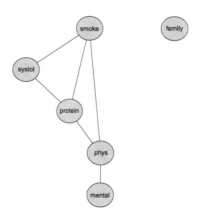

The choice of k is usually argued on the basis of asymptotic considerations. The motivation for AIC is that, under suitable assumptions, it is an approximate measure of the expected Kullback–Leibler distance between the true distribution and the estimated. The BIC difference between two models is the logarithm of a Laplace approximation to the associated Bayes factor for large number of observations n, but a term of lower order of magnitude than $\log n$ is ignored. Under reasonable assumptions the BIC is consistent in the sense that for n large it will choose the simplest model consistent with the data. This will typically only be true for the AIC if that is the saturated model. For a more general discussion of the issues involved, see Ripley (1996, Sect. 2.6).

The default direction is backward but may be changed to forward; notice that we set details=1 to obtain some output from the model selection process:

```
> mildew.init <- dmod(~.^1, data=mildew)
> m.mildew  <- stepwise(mildew.init, k=log(sum(mildew)),
+                          direction="forward", details=1)

STEPWISE:
 criterion: aic ( k = 4.25 )
 direction: forward
 type      : decomposable
 search    : all
 steps     : 1000
. FORWARD: type=decomposable search=all, criterion=aic(4.25),
                                           alpha=0.00
. Initial model: is graphical=TRUE is decomposable=TRUE
  change.AIC  -59.0762 Edge added: a367 p53a
  change.AIC  -55.3386 Edge added: c365 la10
  change.AIC  -48.3388 Edge added: a367 mp58
  change.AIC   -6.3085 Edge added: c365 locc
  change.AIC   -2.1590 Edge added: locc p53a

> plot(m.mildew, "twopi")
```

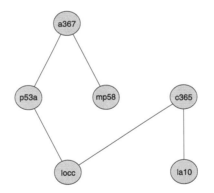

The expression ~.^1 is a shortcut for the main effects model (see Sect. 2.5.2). The selected model shows the order of the markers on the chromosome: see Edwards (2000).

Other variants are possible. Setting headlong=TRUE results in *headlong search*: instead of adding or deleting the edge that gives the greatest decrease in the AIC, the edges at random are examined in random order and the first one found that decreases the AIC is added or deleted. This is generally faster for high-dimensional models.

Output can be suppressed using details=0 whereas setting details=2 will print test statistics for all edges, providings an indication of the strength of evidence for the edges present and the weakness of evidence for the absent edges. When searching among decomposable models (obtained by setting type="decomposable" as opposed to type="unrestricted"), the degrees of freedom are adjusted for sparsity.

```
> mildew.init.2 <- dmod(~.^., data=mildew)
> m.mildew.2  <- stepwise(mildew.init.2, crit="test", alpha=0.05,
+                         details=0)
> m.mildew.2

Model: A dModel with 6 variables
 graphical :  TRUE  decomposable :  TRUE
 -2logL    :        383.01 mdim  :   11 aic :      405.01
 ideviance :        192.46 idf   :    5 bic :      429.74
 deviance  :         17.26 df    :   52
Notice: Table is sparse
  Asymptotic chi2 distribution may be questionable.
  Degrees of freedom can not be trusted.
  Model dimension adjusted for sparsity : 10
```

giving the same model as before.

2.5 Further Topics

2.5.1 Fitting Log-Linear Models with glm()

As we described in Sect. 2.3.5, we could just as well have assumed that cell counts $\{n(i)\}_{i \in \mathcal{I}}$ are independent realisations of Poisson distributions with means $\{\lambda(i)\}_{i \in \mathcal{I}}$. It follows that we can fit log-linear models as generalized linear models by means of the glm() function, using the Poisson distribution and (default) log-link. The estimation method is then Fisher Scoring (which requires inversion of a potentially large matrix).

It is worth mentioning that there may be computational problems with this approach: if the data are sparse and there are only few observations relative to the complexity of the model then the glm() estimation algorithm may fail, as it implicitly assumes that $\hat{m}(i) > 0$ for all $i \in \mathcal{I}$. The IPS algorithm, on the other hand, always works.

The data need to be in aggregrated case list form, as described in Sect. 2.2.2. In the present case we use

```
> lizardAGG

  diam height species Freq
1  <=4  >4.75   anoli   32
2   >4  >4.75   anoli   11
3  <=4 <=4.75   anoli   86
4   >4 <=4.75   anoli   35
5  <=4  >4.75    dist   61
6   >4  >4.75    dist   41
7  <=4 <=4.75    dist   73
8   >4 <=4.75    dist   70
```

We use the Freq variable as response variable. Note that it is important that all cells, also any empty ones, are present in the data. To fit the model shown in (2.1) we can use the code:

```
> m1glm <- glm(Freq~-1+diam:species+height:species,family=poisson,
+                 data=lizardAGG)
> summary(m1glm)

Call:
glm(formula = Freq ~ -1 + diam:species + height:species,
    family = poisson, data = lizardAGG)

Deviance Residuals:
     1        2        3        4        5        6        7        8
 0.190   -0.310   -0.114    0.181    0.687   -0.782   -0.596    0.639

Coefficients:
                         Estimate Std. Error z value Pr(>|z|)
diam<=4:speciesanoli        4.467      0.103   43.30  < 2e-16 ***
diam>4:speciesanoli         3.525      0.155   22.80  < 2e-16 ***
diam<=4:speciesdist         4.359      0.102   42.80  < 2e-16 ***
diam>4:speciesdist          4.171      0.109   38.20  < 2e-16 ***
speciesanoli:height>4.75   -1.035      0.178   -5.83  5.6e-09 ***
speciesdist:height>4.75    -0.338      0.130   -2.61   0.0091 **
---
Signif. codes:  0 '***' 0.001 '**' 0.01 '*' 0.05 '.' 0.1 ' ' 1

(Dispersion parameter for poisson family taken to be 1)

    Null deviance: 2514.8188  on 8  degrees of freedom
Residual deviance:    2.0256  on 2  degrees of freedom
AIC: 59

Number of Fisher Scoring iterations: 4
```

By using glm() we automatically get the asymptotic standard errors of the parameter estimates and also these are not affected by the sampling scheme and hence are valid under both the Poission and multinomial sampling schemes.

By including −1 in the right-hand side of the model formula we set the intercept to zero. This only affects the parametrisation of the model. The residual deviance gives the likelihood ratio test against the saturated model.

```
> msat  <- glm(Freq ~ -1 + diam*height*species, family=poisson,
+                 data=lizardAGG)
> mno3f <- glm(Freq ~ -1 + diam*height + diam*species + species*height,
+                 family=poisson, data=lizardAGG)
> anova(msat, mno3f, m1glm, test="Chisq")

Analysis of Deviance Table

Model 1: Freq ~ -1 + diam * height * species
Model 2: Freq ~ -1 + diam * height + diam * species + species * height
Model 3: Freq ~ -1 + diam:species + height:species
  Resid. Df Resid. Dev Df Deviance P(>|Chi|)
1         0      0.000
2         1      0.149 -1   -0.149      0.70
3         2      2.026 -1   -1.876      0.17
```

Omission of empty cells from the input data corresponds to treating them as structural zeroes. This allows exotic hypotheses such as quasi-independence to be examined (Bishop et al. 1975). But for sparse tables, the `glm()` approach runs into problems, and IPS is to be preferred. For example,

```
> glm(Freq ~.^3, ,family=poisson, data=as.data.frame(mildew))
```

fails to converge but

```
> dmod(~.^3, data=mildew)
```

is unproblematic.

2.5.2 *Working with* `dModel` *Objects*

The `dmod()` function supports some useful shortcut expressions for model formulae. For example, `~.^.` is the saturated model, `~.^1` is the main effect model and `~.^p` is the model with all *p*-factor interactions. Furthermore, to specify marginal models (that is, not including all the variables in the table), the `marginal` argument can be used. Lastly, it is possible to abbreviate variable names. For example,

```
> m <- dmod(~.^2, marginal=c("smo", "prot", "sys","fam"),
+            data=reinis)
```

```
Model: A dModel with 4 variables
 graphical : FALSE  decomposable : FALSE
 -2logL    :        9021.61 mdim :   10 aic :        9041.61
 ideviance :          48.67 idf  :    6 bic :        9096.79
 deviance  :           9.24 df   :    5
```

The generating class of the model as a list and as a right-hand sided formula can be retrieved using `terms()` and `formula()`:

```
> str(terms(m))
```

```
List of 6
 $ : chr [1:2] "smoke" "protein"
 $ : chr [1:2] "smoke" "systol"
 $ : chr [1:2] "smoke" "family"
 $ : chr [1:2] "protein" "systol"
 $ : chr [1:2] "protein" "family"
 $ : chr [1:2] "systol" "family"
```

```
> formula(m)
```

```
~smoke * protein + smoke * systol + smoke * family + protein *
    systol + protein * family + systol * family
```

The dependence graph and adjacency matrix of a model object can be obtained using the `as()` function:

```
> as(m, "graphNEL")
```

```
A graphNEL graph with undirected edges
Number of Nodes = 4
Number of Edges = 6
```

```
> as(m, "matrix")
```

```
        smoke protein systol family
smoke     0       1      1      1
protein   1       0      1      1
systol    1       1      0      1
family    1       1      1      0
```

The update() function enables dModel objects to be modified by the addition or deletion of interaction terms or edges, using the arguments aterm, dterm, aedge or dedge. No prize to work out which does which. Some examples follow:

- Set a marginal saturated model:

```
> ms <- dmod(~.^., marginal=c("phys","mental","systol","family"),
+            data=reinis)
> formula(ms)


~phys * mental * systol * family
```

- Delete one edge:

```
> ms1 <- update(ms, list(dedge=~phys:mental))
> formula(ms1)


~phys * systol * family + mental * systol * family
```

- Delete two edges:

```
> ms2<- update(ms, list(dedge=~phys:mental+systol:family))
> formula(ms2)


~phys * systol + phys * family + mental * systol + mental * family
```

- Delete all edges in a set:

```
> ms3 <- update(ms, list(dedge=~phys:mental:systol))
> formula(ms3)


~phys * family + mental * family + systol * family
```

- Delete an interaction term

```
> ms4 <- update(ms, list(dterm=~phys:mental:systol) )
> formula(ms4)


~phys * mental * family + phys * systol * family + mental * systol *
    family
```

- Add three interaction terms:

```
> ms5 <- update(ms, list(aterm=~phys:mental+phys:systol
                                +mental:systol) )
> formula(ms5)


~phys * mental * systol * family
```

- Add two edges:

```
> ms6 <- update(ms, list(aedge=~phys:mental+systol:family))
> formula(ms6)


~phys * mental * systol * family
```

A brief explanation of these operations may be helpful. To obtain a hierarchical model when we delete a term from a model, we must delete any higher-order relatives to the term. Similarly, when we add an interaction term we must also add all lower-order relatives that were not already present. Deletion of an edge is equivalent to deleting the corresponding two-factor term. Let $m - e$ be the result of deleting edge e from a model m. Then the result of adding e is defined as the maximal model m^* for which $m^* - e = m$.

2.6 Various

Other R packages which support discrete graphical models include **CoCo** (Badsberg 1991) and **gRapHD**, see Chap. 7. The packages **SIN**, **pcalg** and **bnlearn** support algorithms to select discrete graphical models: Sects. 4.4.4, 4.6.1, 4.6.2 and the following sections, illustrate their use in a Gaussian context. Chapter 3 describes the use of discrete, directed graphical models and Sect. 3.4 illustrates the selection of such a model.

Chapter 3
Bayesian Networks

3.1 Introduction

A Bayesian network is traditionally understood to be a graphical model based on a directed acyclic graph (a DAG). The term refers to its use for Bayesian inference in expert systems, where appropriate use of conditional independencies enable rapid and efficient computation of updated probabilities for states of unobserved variables, a calculation which in principle is forbiddingly complex. The term is also used in contrast to the once fashionable neural networks which used quite a different inference mechanism. In principle there is nothing Bayesian about a Bayesian network.

It should be pointed out that the DAG is only used to give a simple and transparent way of specifying a probability model, whereas the simplification in the computations are based on exploiting conditional independencies in an undirected graph. Thus, as we shall illustrate, methods for building undirected graphical models can just as easily be used for building probabilistic inference machines.

The **gRain** package (gRaphical independence network) is an R implementation of such networks. The package implements the propagation algorithm described in Lauritzen and Spiegelhalter (1988). Most of the exposition here is based on the package **gRain**, but **RHugin** is also used, see below. The networks in **gRain** are restricted to discrete variables, each with a finite state space. The package has a similar functionality to that of the GRAPPA suite of functions (Green 2005).

The package **RHugin** provides an R-interface to the (commercial) HUGIN software, enabling access to the full functionality of HUGIN through R. **RHugin** is not on CRAN but is available from http://rhugin.r-forge.r-project.org/. **RHugin** requires a version of HUGIN to be pre-installed. The examples in this chapter which use **RHugin** work with the free version HUGIN Lite, which has full functionality but has size limitations on the networks.

S. Højsgaard et al., *Graphical Models with R*, Use R!,
DOI 10.1007/978-1-4614-2299-0_3, © Springer Science+Business Media, LLC 2012

Fig. 3.1 The directed acyclic
graph corresponding to the
chest clinic example from
Lauritzen and Spiegelhalter
(1988). The arrows indicate a
formalization of the
relationships expressed in the
narrative

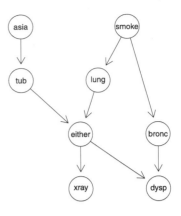

3.1.1 The Chest Clinic Example

This section reviews the chest clinic example of Lauritzen and Spiegelhalter (1988)
(illustrated in Fig. 3.1) and shows one way of specifying a network in **gRain**. Details
of the steps will be given in later sections. Other ways of specifying a network are
described in Sect. 3.3.1. Lauritzen and Spiegelhalter (1988) motivate the example
with the following narrative:

> Shortness-of-breath (dyspnoea) may be due to tuberculosis, lung cancer or bronchitis, or
> none of them, or more than one of them. A recent visit to Asia increases the chances of tu-
> berculosis, while smoking is known to be a risk factor for both lung cancer and bronchitis.
> The results of a single chest X-ray do not discriminate between lung cancer and tuberculo-
> sis, as neither does the presence or absence of dyspnoea.

This narrative is represented by the directed acyclic graph in Fig. 3.1 which forms
the basis for the Bayesian network constructed in this example.

3.1.2 Models Based on Directed Acyclic Graphs

We focus on Bayesian networks for discrete variables and we shall, in accordance
with Chap. 2, use the following notation: Let $X = X_V = (X_v; v \in V)$ be a discrete
random vector. The labels of X_v are generically denoted by i_v so the levels of X are
denoted $i = i_V = (i_v, v \in V)$ and the set of possible values of X is denoted $\mathcal{I}$.

The multivariate distribution associated with a Bayesian network is constructed
by combining univariate (conditional) distributions using the structure of the directed acyclic graph (DAG) with vertices V. To be precise, probability distributions
$p(i_V)$ *factorizes* w.r.t. a directed acyclic graph if it can be expressed as

$$p(i_V) = \prod_{v \in V} p(i_v \mid i_{\mathrm{pa}(v)}) \tag{3.1}$$

i.e. if the joint density or probability mass function is a product of conditional densities of individual variables given their parents in the DAG, see also Sect. 1.3.

For the chest clinic example, write the variables as A = Asia, S = smoker, T = tuberculosis, L = lung cancer, B = bronchitis, D = dyspnoea, X = X-ray and E = either tuberculosis or lung cancer. Each variable can take the values "yes" and "no". Note that E is a logical variable which is true ("yes") if either T or L are true ("yes") and false ("no") otherwise. The DAG in Fig. 3.1 now corresponds to a factorization of the joint probability function $p(i_V)$, where $V = \{A, S, T, L, B, E, D, X\}$ (apologies for using X with two different meanings here) as

$$p(i_A)p(i_S)p(i_T | i_A)p(i_L | i_S)p(i_B | i_S)p(i_E | i_T, i_L)p(i_D | i_E, i_B)p(i_X | i_E). \quad (3.2)$$

In **gRain**, each conditional distribution in (3.2) is specified as a table called a conditional probability table or a CPT for short.

Distributions given as in (3.1) automatically satisfy the *global directed Markov property* so that whenever two sets of nodes A and B are d-separated by a set of nodes S, see Sect. 1.3 for this notion, then $A \perp\!\!\!\perp B \mid S$.

The directed acyclic graph in Fig. 3.1 can be specified as:

```
> g<-list(~asia, ~tub | asia, ~smoke, ~lung | smoke, ~bronc | smoke,
+        ~either | lung : tub, ~xray | either, ~dysp | bronc : either)
> chestdag<-dagList(g)
```

We can query conditional independences using the function d.separates() constructed in Sect. 1.3:

```
> d.separates("tub", "smoke", c("dysp","xray"), chestdag)
```

```
[1] FALSE
```

whereas

```
> d.separates("tub", "lung", "smoke", chestdag)
```

```
[1] TRUE
```

3.1.3 Inference

Suppose we are given evidence that a set of variables $E \subset V$ have a specific value i_E^*. For the chest clinic example, evidence could be that a person has recently visited Asia and suffers from dyspnoea, i.e. i_A = yes and i_D = yes.

With this evidence, we may be interested in the conditional distribution $p(i_v | X_E = i_E^*)$ (or $p(i_v | i_E^*)$ is short) for some of the variables $v \in V \setminus E$ or in $p(i_U | i_E^*)$ for a set $U \subset V \setminus E$. In the chest clinic example, interest might be in $p(i_L | i_E^*)$, $p(i_T | i_E^*)$ and $p(i_B | i_E^*)$, or possibly in the joint (conditional) distribution $p(i_L, i_T, i_B | i_E^*)$. Interest might also be in calculating the probability of a specific event, e.g. $p(i_E^*) = p(X_E = i_E^*)$.

As noticed above, each conditional distribution in (3.2) is in **gRain** specified as a conditional probability table. A brute force approach to calculating $p(i_U | i_E^*)$ is to calculate the joint distribution given by (3.2) by multiplying the conditional probability tables. Finding $p(i_U | i_E^*)$ then reduces to first finding the slice defined by $i_E = i_E^*$ of the joint table and then marginalizing over the variables not in U that slice.

As all variables in the chest clinic example are binary, the joint distribution will have $2^8 = 256$ states but for larger networks/more levels of the variables the joint state space becomes prohibitively large. In most practical cases the set U will be much smaller than V (U might consist of only one or two variables while V can be very large). Combined with the observation that the factorization in (3.2) implies conditional independence restrictions, this implies that $p(i_U \mid i_E^*)$ can be found without ever actually calculating the joint distribution. See Sect. 3.2.3 for details.

3.2 Building and Using Bayesian Networks

3.2.1 Specification of Conditional Probability Tables

One simple way of specifying a model for the chest clinic example is as follows. First we specify conditional probability tables with values as given in Lauritzen and Spiegelhalter (1988). This can be done with `array()` or as here with the `cptable()` function, which offers some additional features:

```
> library(gRain)
> yn <- c("yes","no")
> a    <- cptable(~asia, values=c(1,99), levels=yn)
> t.a  <- cptable(~tub+asia, values=c(5,95,1,99), levels=yn)
> s    <- cptable(~smoke, values=c(5,5), levels=yn)
> l.s  <- cptable(~lung+smoke, values=c(1,9,1,99), levels=yn)
> b.s  <- cptable(~bronc+smoke, values=c(6,4,3,7), levels=yn)
> e.lt <- cptable(~either+lung+tub,values=c(1,0,1,0,1,0,0,1),
                  levels=yn)
> x.e  <- cptable(~xray+either, values=c(98,2,5,95), levels=yn)
> d.be <- cptable(~dysp+bronc+either, values=c(9,1,7,3,8,2,1,9),
                  levels=yn)
```

Notice that the "+" operator used above is slightly misleading in the sense, for example, that the operator does not commute (the order of the variables is important). We use the "+" operator merely as a separator of the variables. The following forms are also valid specifications:

```
> cptable(~tub|asia, values=c(5,95,1,99), levels=yn)
> cptable(c("tub","asia"), values=c(5,95,1,99), levels=yn)
```

Notice that since E is a logical variable which is true if either T or L are true and false otherwise, the corresponding CPT can be created with the special function `ortable()` (there is also an corresponding `andtable()` function):

```
> e.lt <- ortable(~either+lung+tub, levels=yn)
```

3.2.2 Building the Network

A network is created with the function grain() which returns an object of class grain:

```
> plist <- compileCPT(list(a, t.a, s, l.s, b.s, e.lt, x.e, d.be))
> grn1  <- grain(plist)
> summary(grn1)

Independence network: Compiled: FALSE Propagated: FALSE
 Nodes : chr [1:8] "asia" "tub" "smoke" "lung" "bronc" ...

> plot(grn1)
```

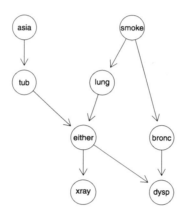

The compileCPT() function does some checking of the specified CPT's. (For example, it is checked that the graph defined by the CPT's is acyclic. Furthermore, the specification of t.a gives a table with four entries and the variable tub is specified to be binary. Hence it is checked that the variable asia is also binary.) The object plist is a list of arrays and it is from this list that the grain object is created.

3.2.2.1 Compilation—Finding the Clique Potentials

A grain object must be compiled and propagated before queries can be made. These steps are performed by the querygrain() function if necessary, but for some purposes it is advantageous to perform them explicitly. Compilation of a network is done with the compile() method for grain objects:

```
> grn1c <- compile(grn1)
> summary(grn1c)

Independence network: Compiled: TRUE Propagated: FALSE
 Nodes : chr [1:8] "asia" "tub" "smoke" "lung" "bronc" ...
 Number of cliques:                    6
 Maximal clique size:                  3
 Maximal state space in cliques:       8
```

Compilation of a `grain` object based on CPTs involves the following steps: First it is checked whether the list of CPTs defines a directed acyclic graph (a DAG). If so, then the DAG is created; it is *moralized* and *triangulated* to form a chordal (triangulated) graph. The CPTs are transformed into *clique potentials* defined on the cliques of this chordal graph. The chordal graph together with the corresponding clique potentials are the most essential components of a `grain` object, and one may indeed construct a `grain` object directly from a specification of these two components, see Sect. 3.3.1.

We again consider the Bayesian network of Sect. 3.2.1: The factorization (3.2) into a *clique potential representation* follows by simply noticing that in (3.2) each of the conditional probability tables can be considered a function of the variables it involves. These potentials are simply non-negative functions.

The dependence graph of the Bayesian network is derived from the potentials. For example, the presence of the term $p(x_D | x_E, x_B)$ implies that there must be edges between all pairs in $\{D, E, B\}$. Algorithmically, the dependence graph can be formed from the DAG by moralization: The moral graph of a DAG is obtained by first joining all parents of each node by a line and then dropping the directions on the arrows. For the chest clinic example, the edges between `tub` and `lung`, and between `either` and `bronc` are added.

The next step is to triangulate the dependence graph if it is not already so by adding additional edges, so-called fill-ins. This is done to enable simple computation of marginals from the clique potentials, cf. Sect. 3.2.2.2 below. Finding an optimal triangulation (in terms of a minimal number of fill-ins) of a given graph is NP-complete, but various good heuristics exist. The **gRbase** package implements a Minimum Clique Weight Heuristic method inspired by Kjærulff (1990). Two possible fill-ins are the edge between `lung` and `bronc`, and the edge between `either` and `smoke`. The triangulated graph is also a dependence graph for (3.2); the graph just conceals some conditional independence restrictions implied by the model.

The steps described above can alternatively be carried out separately, and Fig. 3.2 illustrates the process:

```
> g   <- grn1$dag
> mg  <- moralize(g)
> tmg <- triangulate(mg)
```

Recall from Sect. 1.2.1 that an ordering $C_1, \ldots, C_T$ of the cliques of a graph is a RIP ordering if $S_j = (C_1 \cup \cdots \cup C_{j-1}) \cap C_j$ is contained in one (but possibly several) of the cliques $C_1, \ldots, C_{j-1}$, obtained with:

```
> rip(tmg)

cliques
  1 : tub asia
  2 : either tub lung
  3 : bronc lung either
  4 : smoke lung bronc
  5 : dysp bronc either
  6 : xray either
```

mg   **tmg**

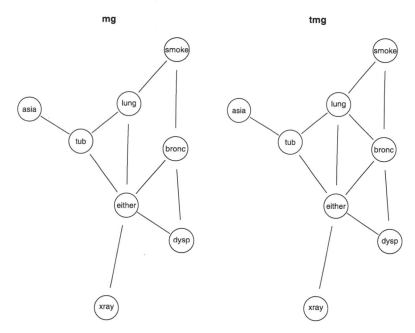

Fig. 3.2 *Left*: moralized DAG; *Right*: triangulated moralized DAG. The chect clinic example of Lauritzen and Spiegelhalter (1988)

```
separators
  1 :
  2 : tub
  3 : lung either
  4 : lung bronc
  5 : bronc either
  6 : either
parents
  1 : 0
  2 : 1
  3 : 2
  4 : 3
  5 : 3
  6 : 5
```

Picking a particular clique, say C_k, with $S_j \subseteq C_k$ and naming this as the *parent clique* of C_j, with C_j being the child of C_k, organizes the cliques of the triangulated graph in a rooted tree with the cliques as nodes and arrows from parent to child. We call S_j the separator and $R_j = C_j \setminus S_j$ the residual, where $S_1 = \emptyset$. The junction tree is formed by ignoring the root and the directions on the edges. It is a tree with the property that for any pair (A, B) of cliques and any clique C on the unique path between A and B it holds that $A \cap B \subseteq C$. It can be shown that *the cliques of a graph can be organized in a junction tree if and only if the graph is triangulated.*

The junction tree can be displayed by plot(),

```
> plot(grn1c,type="jt")
```

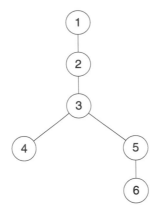

where the numbers on the nodes refer to the clique numbers in the RIP-ordering. Other RIP-orderings of the cliques can be found by choosing an arbitrary clique as the first and then numbering the cliques in any way which is increasing as one moves outward from the first clique in this tree. For example $C_3, C_2, C_5, C_1, C_6, C_4$ would be another RIP-ordering.

The functions $p(i_v \mid i_{\text{pa}(v)})$ are hence defined on complete sets of the triangulated graph. For each clique C we collect the conditional probability tables $p(i_v \mid i_{\text{pa}(v)})$ into a single term $\psi_C(i_C)$ by multiplying them. Triangulation may have created cliques to which no CPT corresponds. For each such clique the corresponding potential is identically equal to 1. Thus we have obtained the clique potential representation of $p(i_V)$ as

$$p(i_V) = \prod_{j=1}^{T} \psi_{C_j}(i_{C_j}). \tag{3.3}$$

The representation (3.3) is the fundamental representation for the subsequent computations. As such, a DAG and a corresponding factorization as in (3.2) is just one way of getting to the representation in (3.3) and one may alternatively specify this directly as shall be illustrated in Sect. 3.3.1.

3.2.2.2 Propagation—from Clique Potentials to Clique Marginals

To be able to answer queries, the `grain` object must be propagated, which means that the clique potentials must be calibrated (adjusted) to each other. Propagation is done with the `propagate()` method for `grain` objects:

```
> grn1c <- propagate(grn1c)
> summary(grn1c)

Independence network: Compiled: TRUE Propagated: TRUE
  Nodes : chr [1:8] "asia" "tub" "smoke" "lung" "bronc" ...
  Number of cliques:                    6
  Maximal clique size:                  3
  Maximal state space in cliques:       8
```

The propagation algorithm works by turning the clique potential representation (3.3) into a representation in which each potential ψ_{C_j} is replaced by the marginal distribution $p(i_{C_j})$. This representation is called a *clique marginal representation*. This is done by working twice through the set of cliques and passing 'messages' between neighbouring cliques: first from the last clique in the RIP-ordering towards the first, i.e. inwards in the junction tree, and subsequently passing messages in the other direction.

In detail, the process is as follows. We start with the last clique C_T in the RIP ordering where $C_T = S_T \cup R_T$, $S_T \cap R_T = \emptyset$. The factorization (3.3) together with the factorization criterion (1.1) implies that $R_T \perp\!\!\!\perp (C_1 \cup \cdots \cup C_{T-1}) \setminus S_T \mid S_T$. Marginalizing over i_{R_T} gives

$$p(i_{C_1 \cup \ldots \cup C_{T-1}}) = \left(\prod_{j=1}^{T-1} \psi_{C_j}(i_{C_j}) \right) \sum_{i_{R_T}} \psi_{C_T}(i_{S_T}, i_{R_T}).$$

Let $\psi_{S_T}(i_{S_T}) = \sum_{i_{R_T}} \psi_{C_T}(i_{S_T}, i_{R_T})$. Then from the expression above we have

$$p(i_{R_T} \mid i_{S_T}) = \psi_{C_T}(i_{S_T}, i_{R_T}) / \psi_{S_T}(i_{S_T})$$

and hence

$$p(i_V) = p(i_{C_1 \cup \ldots \cup C_{T-1}}) p(i_{R_T} \mid i_{S_T}) = \left\{ \left(\prod_{j=1}^{T-1} \psi_{C_j}(i_{C_j}) \right) \psi_{S_T}(i_{S_T}) \right\} \frac{\psi_{C_T}(i_{C_T})}{\psi_{S_T}(i_{S_T})}.$$

The RIP ordering ensures that S_T is contained in the neighbour of C_T in the junction tree (one of the cliques $C_1, \ldots, C_{T-1}$), say C_k. We can therefore absorb ψ_{S_T} into ψ_{C_k} by setting $\psi_{C_k}(i_{C_k}) \leftarrow \psi_{C_k}(i_{C_k}) \psi_{S_T}(i_{S_T})$. We can think of the clique C_T *passing the message* ψ_{S_T} to its neighbour C_k, making a note of this by changing its own potential to $\psi_{C_T} \leftarrow \psi_{C_T} / \psi_{S_T}$, and C_k absorbing the message.

After this we now have $p(i_{C_1 \cup \ldots \cup C_{T-1}}) = \prod_{j=1}^{T-1} \psi_{C_j}(i_{C_j})$. We can then apply the same scheme to the part of the junction tree which has not yet been traversed. Continuing in this way until we reach the root of the junction tree yields

$$p(i_V) = p(i_{C_1}) p(i_{R_2} \mid i_{S_2}) p(i_{R_3} \mid i_{S_3}) \ldots p(i_{R_T} \mid i_{S_T}) \tag{3.4}$$

where $p(i_{C_1}) = \psi_{C_1}(i_{C_1}) / \sum_{i_{C_1}} \psi_{C_1}(i_{C_1})$. The resulting expression (3.4) is called a *set chain representation*. Note that the root potential now yields the joint marginal distribution of its nodes.

For some purposes we do not need to proceed further and the set chain representation is fully satisfactory. However, if we wish to calculate marginals to other cliques than the root clique, we need a second passing of messages. This time we work from the root towards the leaves in the junction tree and send messages with a slight twist, in the sense that this time we do not change the potential in the sending clique. Rather we do as follows:

Suppose C_1 is the parent of C_2 in the rooted junction tree. Then we have that $p(i_{S_2}) = \sum_{i_{C_1 \setminus S_2}} p(i_{C_1})$ and so

$$p(i_V) = p(i_{C_1}) \frac{p(i_{C_2})}{p(i_{S_2})} p(i_{R_3} \mid i_{S_3}) \ldots p(i_{R_T} \mid i_{S_T}).$$

Thus when the clique C_2 has absorbed its message by the operation

$$\psi_{C_2}(i_{C_2}) \leftarrow \psi_{C_2}(i_{C_2})p(i_{S_2})$$

its potential is equal to the marginal distribution of its nodes. Proceeding in this way until we reach the leaves of the junction tree yields the clique marginal representation

$$p(i_V) = \prod_{j=1}^{T} p(i_{C_j}) / \prod_{j=2}^{T} p(i_{S_j}). \tag{3.5}$$

3.2.3 Absorbing Evidence and Answering Queries

Consider absorbing the evidence $i_E^* = (i_v^*, v \in E)$, i.e. that nodes in E are known to have a specific value. We note that

$$p(i_{V \setminus E} \mid i_E^*) \propto p(i_{V \setminus E}, i_E^*)$$

and hence evidence can be absorbed into the model by modifying the terms ψ_{C_j} in the clique potential representation (3.3) as follows: for every $v \in E$, we take an arbitrary clique C_j with $v \in C_j$. Entries in ψ_{C_j} which are locally inconsistent with the evidence, i.e. entries i_{C_j} for which $i_v \neq i_v^*$, are set to zero and all other entries are unchanged. Evidence can be entered before or after propagation without changing final results.

To answer queries, we carry out the propagation steps above leading to a clique marginal representation where the potentials become $\psi_{C_j}(i_{C_j}) = p(i_{C_j} \mid i_E^*)$. In this process we note that processing of the root potential to find $p(i_{C_1} \mid i_E^*)$ involves computation of $\sum_{i_{C_1}} \psi_1(i_{C_1})$ which is equal to $p(i_E^*)$. Hence the probability of the evidence comes at no extra computational cost.

Evidence is entered with setFinding() which creates a new grain object:

```
> grn1c.ev <-
+ setFinding(grn1c,nodes=c("asia","dysp"), states=c("yes","yes"))
```

To obtain $p(i_v \mid i_E^*)$ for some $v \in V \setminus E$, we locate a clique C_j containing v and marginalize as $\sum_{i_{C_j \setminus \{v\}}} p(i_{C_j})$. Based on (3.5) the grain objects with and without evidence can now be queried to give marginal probabilities using querygrain():

```
> querygrain(grn1c.ev,nodes=c("lung","bronc"), type="marginal")

$lung
lung
    yes      no
0.09953 0.90047

$bronc
bronc
   yes     no
0.8114 0.1886
```

```
> querygrain(grn1c,nodes=c("lung","bronc"), type="marginal")

$lung
lung
   yes     no
0.055 0.945

$bronc
bronc
 yes   no
0.45 0.55
```

The evidence in a grain object can be retrieved with the getFinding() function while the probability of observing the evidence is obtained using the pFinding() function:

```
> getFinding(grn1c.ev)

Finding:
      variable state
[1,] asia      yes
[2,] dysp      yes
Pr(Finding)= 0.004501

> pFinding(grn1c.ev)

[1] 0.004501
```

Suppose we want the distribution $p(i_U \mid i_E^*)$ for a set $U \subset V \setminus E$. If there is a clique C_j such that $U \subset C_j$ then the distribution is simple to find by summing $p(i_{C_j})$ over the variables in $C_j \setminus U$. If no such clique exists we can obtain $p(i_U \mid i_E^*)$ by calculating $p(i_U^*, i_E^*)$ for all possible configurations i_U^* of U and then normalizing the result: this can be computationally demanding if U has a large state space.

```
> querygrain(grn1c.ev,nodes=c("lung","bronc"), type="joint")

      bronc
lung       yes       no
  yes 0.06298 0.03654
  no  0.74842 0.15205
> querygrain(grn1c.ev,nodes=c("lung","bronc"), type="conditional")

      bronc
lung      yes      no
  yes 0.07762 0.1938
  no  0.92238 0.8062
```

Note that the latter result is the conditional distribution of lung given bronc—but also conditional on the evidence.

However, if it is known beforehand that interest will be in the joint distribution of a specific set U of variables, one can ensure that the set U is contained in a single clique in the triangulated graph. This can for example be done by first moralizing, then adding edges between all nodes in U, and finally triangulating the resulting graph. The price to be paid is that the cliques may become larger and since computational complexity is exponential in the largest clique, this can be prohibitive.

To do this in practice we first need to compile the grain again

```
> grn1c2 <- compile(grn1, root=c("lung", "bronc", "tub"),
                     propagate=TRUE)
> grn1c2.ev  <- setFinding(grn1c2,nodes=c("asia","dysp"),
+                              states=c("yes","yes"))
```

Now compare the computing times: the second method is much faster:

```
> system.time({for (i in 1:50)
+                 querygrain(grn1c.ev,nodes=c("lung","bronc","tub"),
+                              type="joint")
+              })

   user  system elapsed
    1.5     0.0     1.5

> system.time({for (i in 1:50)
+                 querygrain(grn1c2.ev,nodes=c("lung","bronc","tub"),
+                              type="joint")
+              })

   user  system elapsed
   0.02    0.00    0.01
```

Evidence can be entered incrementally by calling setFinding() repeatedly. It is most efficient to set propagate=FALSE in setFinding() and then only call the propagate() method for grain objects at the end:

```
> grn1c.ev <- setFinding(grn1c,nodes="asia", states="yes",
+                          propagate=FALSE)
> grn1c.ev <- setFinding(grn1c.ev,nodes="dysp", states="yes",
+                          propagate=FALSE)
> grn1c.ev <- propagate(grn1c.ev)
```

Evidence can be retracted (removed) using the retractFinding() function:

```
> grn1c.ev <- retractFinding(grn1c.ev, nodes="asia")
> getFinding(grn1c.ev)

Finding:
     variable state
[1,] dysp     yes
Pr(Finding)= 0.004501
```

Omitting nodes implies that all evidence is retracted, i.e. that the grain object is reset to its original status.

3.3   Further Topics

3.3.1   Building a Network from Data

A grain object can also be built from a dataframe of cases in various ways, as illustrated below.

One way is to build it is to use data in combination with a graph such as, for example, the directed acyclic graph chestdag specified in Sect. 3.1.2.

The data chestSim500 from the **gRbase** package is generated from our fictitious example using the command simulate() method described in Sect. 3.3.3 below.

When building a grain object this way, the CPTs are estimated from data in chestSim500 as the relative frequencies. To avoid zeros in the CPTs one can choose to add a small number, e.g. smooth=0.1 to all values, corresponding to a Bayesian estimate based on prior Dirichlet distributions for the CPT entries:

```
> library(gRbase)
> data(chestSim500, package='gRbase')
> simdagchest <- grain(chestdag, data=chestSim500)
> simdagchest <- compile(simdagchest, propagate=TRUE, smooth=.1)
> querygrain(simdagchest, nodes =c("lung","bronc"),type="marginal")

$lung
lung
  yes    no
0.046 0.954

$bronc
bronc
  yes    no
0.454 0.546
```

Alternatively, a grain object can be built from an undirected (but triangulated) graph rather than a Bayesian network, making some steps of the process of compilation redundant. The undirected triangulated graph for the compiled chest clinic example can be specified as:

```
> g<-list(~asia : tub, ~either : lung : tub, ~either : lung : smoke,
+     ~bronc : either : smoke, ~bronc : dysp : either, ~either :
+          xray)
> myug <- ugList(g)
```

A grain object can now be built from the graph and the data. In this process, the clique potentials are directly estimated by the appropriate frequencies:

```
> simugchest <- grain(myug, data=chestSim500)
> simugchest <- compile(simugchest, propagate=TRUE)
> plot(simugchest)
```

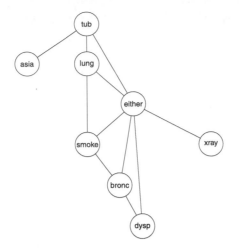

This is natural when directions are not known beforehand. For example, using the
`reinis` data with a model selection procedure yields

```
> data(reinis, package='gRbase')
> m0 <- dmod(~.^., data=reinis)
> m1 <- stepwise(m0)
> reinisgrain <- grain(as(m1,"graphNEL"), data=reinis)
> plot(reinisgrain)
> reinisgrain <- compile(reinisgrain, propagate=TRUE)
> querygrain(reinisgrain,nodes=c("phys","protein"), type="marginal")
```

```
$protein
protein
      y       n
0.5763 0.4237

$phys
phys
      y       n
0.5035 0.4965
```

Now evidence can be entered and revised probabilities found as usual:

```
> reinisgrain.ev  <-
+ setFinding(reinisgrain,
+              nodes=c("systol","smoke","mental"), states=c("y","y","y"))
> querygrain(reinisgrain.ev,nodes=c("phys","protein"), type="marginal")

$protein
protein
      y      n
0.6744 0.3256

$phys
phys
      y      n
0.2776 0.7224
```

3.3.2 Bayesian Networks with RHugin

The package **RHugin** (see http://rhugin.r-forge.r-project.org) provides an Application Programmer's Interface (API) to HUGIN in the R language. It consists of a basic library of functions which mirrors the C API

   http://www.hugin.com/developer/documentation/API_Manuals/

provided with HUGIN. More precisely, every command in the C API of HUGIN of the form h_something has an R-variant called RHugin_something, e.g. RHugin_domain_propagate uses .Call to invoke the HUGIN function h_domain_propagate etc. In this way, the full functionality of HUGIN becomes available within R.

   In addition, **RHugin** provides a few higher level functions based on this API which enables simple operations for Bayesian networks to be carried out, for example such as those described in the previous sections. For the first simple illustrations we repeat the basic steps above using **RHugin** instead of **gRain**.

   We first create the domain

```
> library(RHugin)
> RHchestClinic <- hugin.domain()
```

and subsequently create nodes and give them states

```
> chestNames <- c("asia", "smoke", "tub", "lung", "bronc",
+                 "either", "xray", "dysp")
> for(node in chestNames)
+   add.node(RHchestClinic, node, states = c("yes", "no"))
```

Then nodes are connected with edges to form the DAG

```
> add.edge(RHchestClinic, "tub", "asia")
> add.edge(RHchestClinic, "lung", "smoke")
> add.edge(RHchestClinic, "bronc", "smoke")
> add.edge(RHchestClinic, "either", c("tub", "lung"))
> add.edge(RHchestClinic, "xray", "either")
> add.edge(RHchestClinic, "dysp", c("either", "bronc"))
```

The network now exists and can be displayed using **Rgraphviz**

```
> library(Rgraphviz)
> plot(RHchestClinic)
```

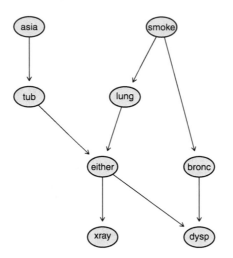

At this point the network has default (uniform) probability tables:

```
> get.table(RHchestClinic, "dysp")
```

```
  dysp either bronc Freq
1  yes    yes   yes    1
2   no    yes   yes    1
3  yes     no   yes    1
4   no     no   yes    1
5  yes    yes    no    1
6   no    yes    no    1
7  yes     no    no    1
8   no     no    no    1
```

These can now be changed manually:

```
> cpt <- get.table(RHchestClinic, "asia")
> cpt$Freq <- c(0.01, 0.99)
> set.table(RHchestClinic, "asia", cpt)
> cpt <- get.table(RHchestClinic, "tub")
> cpt$Freq <- c(5, 95, 1, 99)
> set.table(RHchestClinic, "tub", cpt)
> cpt <- get.table(RHchestClinic, "either")
> cpt
```

```
  either tub lung Freq
1    yes yes  yes    1
2     no yes  yes    1
3    yes  no  yes    1
4     no  no  yes    1
5    yes yes   no    1
6     no yes   no    1
```

```
7     yes  no   no    1
8      no  no   no    1
> cpt$Freq <- c(1,0,1,0,1,0,0,1)
> set.table(RHchestClinic, "either", cpt)
>
```

or using available data to populate one of the tables:

```
> set.table(RHchestClinic,"dysp",chestSim500)
```

leading to

```
> get.table(RHchestClinic, "dysp")
  dysp either bronc Freq
1  yes    yes   yes   10
2   no    yes   yes    2
3  yes     no   yes  176
4   no     no   yes   39
5  yes    yes    no   12
6   no    yes    no    5
7  yes     no    no   29
8   no     no    no  227
```

Note that the CPTs are not yet normalized. In HUGIN this happens at the stage of compilation. We can also let most (or all) tables be based on frequencies in the dataframe:

```
> set.table(RHchestClinic, "smoke", chestSim500)
> set.table(RHchestClinic, "bronc", chestSim500)
> set.table(RHchestClinic, "lung", chestSim500)
> set.table(RHchestClinic, "xray", chestSim500)
> get.table(RHchestClinic, "smoke")

  smoke Freq
1   yes  238
2    no  262
```

If we compile the network we find that tables have become normalized:

```
> compile(RHchestClinic)
> get.table(RHchestClinic, "dysp")

  dysp either bronc   Freq
1  yes    yes   yes 0.8333
2   no    yes   yes 0.1667
3  yes     no   yes 0.8186
4   no     no   yes 0.1814
5  yes    yes    no 0.7059
6   no    yes    no 0.2941
7  yes     no    no 0.1133
8   no     no    no 0.8867
```

The network is now ready for absorbing evidence and calculating revised probabilities:

```
> set.finding(RHchestClinic, "asia","yes")
> set.finding(RHchestClinic, "dysp","yes")
> propagate(RHchestClinic)
> get.belief(RHchestClinic, "lung")
```

```
    yes       no
0.07729 0.92271

> get.belief(RHchestClinic, "bronc")

  yes    no
0.806 0.194
```

Note the values are somewhat different from those previously obtained. This is due
to the fact that probabilities are estimated from the (simulated) data rather than
specified exactly.

3.3.3 Simulation

It is possible to simulate data from a Bayesian network model. The methods use
the current clique potentials to do this and thus generates values conditional on all
evidence entered in the grain object. It uses the method of random propagation as
described in Dawid (1992); see also Cowell et al. (1999, p. 99). If a domain is not
propagated when simulate() is applied, simulate() will force this to happen
automatically.

```
> simulate(grn1c.ev, nsim=5)

  asia tub smoke lung bronc either xray dysp
1  yes  no   yes   no   yes     no   no  yes
2  yes yes   yes   no   yes    yes  yes  yes
3  yes  no   yes   no    no     no   no  yes
4  yes  no    no   no   yes     no   no  yes
5  yes  no    no   no   yes     no   no  yes
```

One application of such a simulation is to obtain the joint distribution of lung and
bronc conditional on the finding:

```
> xtabs(~lung+bronc, data=simulate(grn1c.ev, nsim=1000))/1000

      bronc
lung     yes    no
  yes 0.070 0.033
  no  0.757 0.140
```

The result of the simulation is close to the exact result given in Sect. 3.2.3. A simu-
late() method is also available with **RHugin**, but this only works if the domain has
been propagated.

```
> simulate(RHchestClinic, nsim=5)

  asia smoke tub lung bronc either xray dysp
1  yes   yes  no   no   yes     no   no  yes
2  yes    no  no   no   yes     no  yes  yes
3  yes    no  no   no   yes     no   no  yes
4  yes    no  no   no   yes     no   no  yes
5  yes    no  no   no   yes     no   no  yes
```

3.3.4 Prediction

A predict method is available for grain objects for predicting a set of "responses" from a set of "explanatory variables". Two types of predictions can be made. The default is type="class" which assigns the value to the class with the highest probability:

```
> mydata

  bronc dysp either lung tub asia xray smoke
1  yes  yes   yes   yes  no   no  yes   yes
2  yes  yes   yes   yes  no   no  yes    no
3  yes  yes   yes    no yes   no  yes   yes
4  yes  yes    no    no  no  yes  yes    no

> predict(grn1c, response=c("lung","bronc"), newdata=mydata,
+   predictors=c("smoke", "asia", "tub", "dysp", "xray"), type="class")

$pred
$pred$lung
[1] "yes" "no"  "no"  "no"

$pred$bronc
[1] "yes" "yes" "yes" "yes"


$pFinding
[1] 0.0508476 0.0111697 0.0039778 0.0001083
```

The output should be read carefully: Conditional on the first observation in mydata, the most probable value of lung is "yes" and the same is the case for bronc. This is not in general the same as saying that the most likely configuration of the two variables lung and bronc is "yes".

The entire conditional distribution can be obtained in **gRain** by setting type='dist':

```
> predict(grn1c, response=c("lung","bronc"), newdata=mydata,
+   predictors=c("smoke", "asia", "tub", "dysp", "xray"), type="dist")

$pred
$pred$lung
        yes     no
[1,]  0.7745 0.2255
[2,]  0.3268 0.6732
[3,]  0.1000 0.9000
[4,]  0.3268 0.6732

$pred$bronc
        yes     no
[1,]  0.7182 0.2818
[2,]  0.6373 0.3627
[3,]  0.6585 0.3415
[4,]  0.6373 0.3627


$pFinding
[1] 0.0508476 0.0111697 0.0039778 0.0001083
```

The jointly most probably configuration can be found by using the option `equilibrium ="max"` with **RHugin**. HUGIN uses the max-propagation algorithm described in Dawid (1992); see also Cowell et al. (1999, p. 97 ff.). For the third datapoint we get

```
> initialize(RHchestClinic)
```

```
A Hugin domain
Nodes: asia smoke tub lung bronc either xray dysp
Edges:
  asia -> tub
  smoke -> bronc
  smoke -> lung
  tub -> either
  lung -> either
  bronc -> dysp
  either -> dysp
  either -> xray
```

```
> set.finding(RHchestClinic,"smoke","yes")
> set.finding(RHchestClinic,"asia","no")
> set.finding(RHchestClinic,"tub","yes")
> set.finding(RHchestClinic,"dysp","yes")
> set.finding(RHchestClinic,"xray","yes")
```

The joint probability of the evidence is

```
> propagate(RHchestClinic)
> pev<-get.normalization.constant(RHchestClinic)
> pev
```

```
[1] 0.003687
```

and the most likely configuration is

```
> propagate(RHchestClinic,equilibrium ="max")
> get.belief(RHchestClinic,"either")
yes  no
  1   0
```

```
> get.belief(RHchestClinic,"lung")

    yes      no
0.08676 1.00000
```

```
> get.belief(RHchestClinic,"bronc")

   yes      no
1.0000 0.5627
```

The most probable configuration of the unobserved nodes either, lung, bronc is found by combining states where get.belief() returns 1.00, in this case yes, no, yes. The second number indicates how much the joint probability decreases if the state at that particular node is changed, i.e. the joint probability of yes,yes,yes,

is .08676 times the maximal probability. The probability of the most probable configuration and evidence jointly is obtained via the normalization constant again

```
> pmax<-get.normalization.constant(RHchestClinic)
> pmax
```

```
[1] 0.002171
```

So the conditional probability of the most probable configuration given the evidence
is

```
> predprob<-pmax/pev
> predprob
```

```
[1] 0.5888
```

To simulate with **RHugin**, we now need to propagate again with the default "sum"
option.

3.3.5 Working with HUGIN Files

With HUGIN, the specifications of a BN are read or saved in a textfile in a format
known as a .net. HUGIN can also save and read domains in its own binary format
.hkb which can contain further information in the form of compilation, evidence,
and propagation results.

A grain object saved in this format can be loaded into R using the loadHugin-
Net() function in **gRain**:

```
> chest <- loadHuginNet("ChestClinic.net")
```

```
> chest
```

```
Independence network: Compiled: FALSE Propagated: FALSE
 Nodes: chr [1:8] "PositiveXray" "Bronchitis" "Dyspnoea" ...
```

HUGIN distinguishes between node names and node labels. Node names have to be
unique; node labels need not be so. When creating a BN in HUGIN node names
are generated automatically as C1, C2 etc. The user can choose to give more informative labels or to give informative names. Typically one would do the former.
Therefore loadHuginNet uses node labels (if given) from the netfile and otherwise
node names.

This causes two types of problems. First, HUGIN allows spaces and special characters (e.g. "?") in variable labels, but these are not allowed in **gRain**. If such a
name is found by loadHuginNet, it is converted as follows: special characters are
removed, the first letter after a space is capitalized and then spaces are removed.
Hence the label "visit to Asia?" in a net file will be converted to "visitToAsia". The
same convention applies to states of variables. Secondly, because node labels in the

net file are used as node names in **gRain** we may end up with two nodes having the same name, which is obviously not permitted. To resolve this **gRain** will in such cases force the node names in **gRain** to be the node names rather than the node labels from the net file. For example, if nodes A and B in a net file both have label foo, then the nodes in **gRain** will be denoted A and B. Note that this approach is not entirely foolproof: If there is a node C with label A, then we have just moved the problem. So the scheme above is applied recursively until all ambiguities are resolved.

A grain can be saved in the .net format with the saveHuginNet() function.

```
> saveHuginNet(reinisgrain,file="reinisgrain.net")
```

Note that reinisgrain does not have a DAG by default, so the save function constructs a DAG which has the same conditional independence structure as the triangulated graph defining the grain object.

RHugin interacts continuously with HUGIN but has also read and write functions write.rhd() and read.rhd(). For example we can now create a domain in **RHugin** as

```
> RHreinis<-read.rhd("reinisgrain.net")
> RHreinis

A Hugin domain
Nodes: family mental phys systol smoke protein
Edges:
  mental -> family
  phys -> mental
  systol -> protein
  smoke -> mental
  smoke -> phys
  smoke -> protein
  smoke -> systol
  protein -> phys
```

We can now operate fully in RHreinis with **RHugin**, for example

```
> get.table(RHreinis,"mental")
> set.finding(RHreinis,"mental","y")
> set.finding(RHreinis,"protein","n")
> compile(RHreinis)
> propagate(RHreinis)
> get.normalization.constant(RHreinis)
> get.belief(RHreinis,"smoke")
> write.rhd(RHreinis,"RHreinis.hkb",type="hkb")
> write.rhd(RHreinis,"RHreinis.net",type="net")
```

The file RHreinis.hkb will now be in a binary format readable by HUGIN (or **RHugin**) and contain a compiled and propagated domain with its evidence and the associated .net file, whereas RHreinis.net will be a textfile identical to reinisgrain.net. Similarly, **RHugin** can also read domains which are saved in .hkb format.

It is important to emphasize the relationship between **RHugin** and **gRain** on the one side and HUGIN on the other: **gRain** works entirely within R, creates R objects, and only relates to HUGIN through its facilities for reading and writing .net files. In contrast, domains, nodes, edges etc. of an **RHugin**-domain are not R objects as such. They are *external pointers* to objects which otherwise exist within HUGIN. So, for example, a statement of the form

```
> newRHreinis<-RHreinis
```

does not create a new R object, but just an extra pointer to the same HUGIN domain. This means that if anything is changed in RHreinis, it will automatically change in the same way in newRHreinis and vice versa.

3.4 Learning Bayesian Networks

Hitherto in this chapter it has been assumed that the Bayesian network was known in advance. In practice it is often necessary to construct the network based on an observed dataset—a topic known in the machine learning community as *structural learning* (in contrast to *parameter* learning) and in the statistical community as *model selection*.

Model selection algorithms for Gaussian graphical models based on DAGs are described in Chap. 4. Available algorithms include the PC-algorithm (Spirtes and Glymour 1991) and various algorithms in the **bnlearn** package: these can also be used to select discrete Bayesian networks.

As illustration we consider a dataframe cad1 supplied along with the **gRbase** package. This contains data on coronary artery disease from a Danish heart clinic. In all 14 discrete variables were recorded for 236 patients at the clinic including five background variables (sex, hypercholesterolemia, smoking, heridary disposition and workload), one recording whether or not the patient has coronary artery disease, four variables representing disease manifestation (hypertrophy, previous myocardial infarct, angina pectoris, other heartfailures), and four clinical measurements (Q-wave, T-wave, Q-wave informative and T-wave informative). These data were used as an example of chain graph modelling in Højsgaard and Thiesson (1995).

As a first attempt we can apply the hill-climbing algorithm implemented in the hc function in the **bnlearn** package. This is a greedy algorithm to find a model optimizing a score: various scores may be used, and here we choose to minimize the Bayesian Information Criterion (BIC).

```
> library(gRbase)
> data(cad1, package='gRbase')
> library(bnlearn)
> cad.bn <- hc(cad1)
> plot(as(amat(cad.bn), "graphNEL"))
```

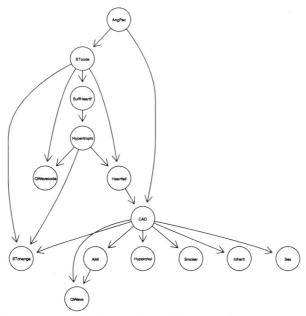

As described in more detail in Sect. 4.5.1, DAGs can only be selected up to Markov equivalence, so it is useful to see which DAGs are Markov equivalent to the selected one. These may be represented as an essential graph, using the `essentialGraph` function in the **ggm** package.

```
> library(ggm)
> plot(as(essentialGraph(amat(cad.bn)), "graphNEL"))
```

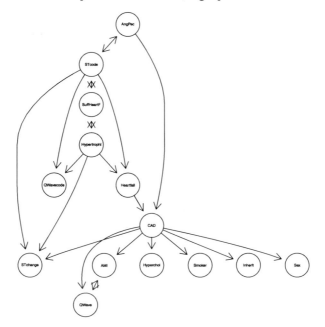

This model is implausible, since it suggests amongst other things that whether or not a patient has coronary artery disease (CAD) determines their sex and whether they smoke. A better approach is to incorporate our prior knowledge of the system under study into the model selection process. We do this by dividing the variables into the four blocks described above, namely background variables, disease, disease manifestations and clinical measurements. Note that we treat hypertrophy as a disease variable, following Højsgaard and Thiesson (1995). We restrict the model selection process by blacklisting (i.e., disallowing) arrows that point from a later to an earlier block. The following code shows how this may be done. First we construct an adjacency matrix containing the disallowed edges, then we convert this into a dataframe using the get.edgelist function in the **igraph** package. This is then passed to the hc function.

```
> block <- c(1,3,3,4,4,4,4,1,2,1,1,1,3,2)
> blM <- matrix(0, nrow=14, ncol=14)
> rownames(blM) <- colnames(blM) <- names(cad1)
> for (b in 2:4) blM[block==b, block<b] <- 1
> library(igraph)
> blackL <- data.frame(get.edgelist(as(blM, "igraph")))
> names(blackL) <- c("from", "to")
> cad.bn1 <- hc(cad1, blacklist=blackL)
> plot(as(amat(cad.bn1), "graphNEL"))
```

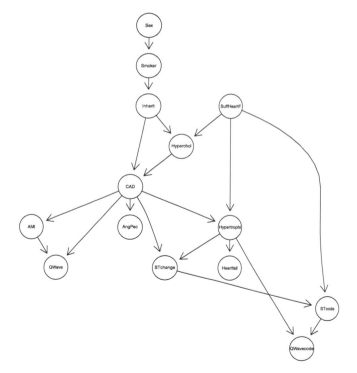

Finally, we again examine the essential graph of the selected DAG:

```
> library(ggm)
> plot(as(essentialGraph(amat(cad.bn1)), "graphNEL"))
```

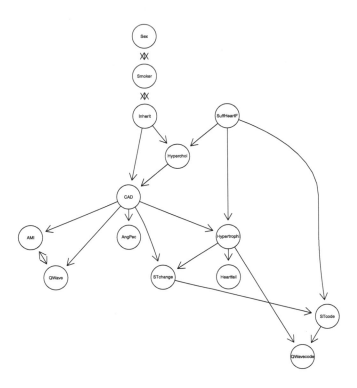

This is more plausible. To create the corresponding grain object, we can use

```
> cad.gr <- as(amat(cad.bn1), "graphNEL")
> cad.grain <- grain(cad.gr, data=cad1)
```

and proceed as before.

Chapter 4
Gaussian Graphical Models

4.1 Introduction

This chapter describes graphical models for multivariate Gaussian data. This is an area which has been under extensive development in recent years, as witnessed by the intensive development of R-packages with facilities for the analysis of such data, for example **ggm**, **deal**, **SIN**, **glasso**, **qp**, **gRc**, **pcalg**, **bnlearn**, and others. We refrain from giving a description of all these packages and focus rather on introducing some fundamental ideas and facilities, illustrated by use of the relevant packages when appropriate.

Section 4.2 introduces the topic via an illustrative dataset. Section 4.3 gives a brief theoretical exposition of undirected graphical models for Gaussian data, and the following section describes some model selection algorithms for these models. The next two sections describe corresponding models based on directed graphs (DAGs) and some model selection algorithms for these. The final section gives a brief description of models based on chain graphs, and a selection algorithm for these.

4.2 Some Examples

4.2.1 Carcass Data

The `carcass` data from **gRbase** contains measurements of the thickness of meat and fat layers at different locations on the back of a slaughter pig together with the lean meat percentage on each of 344 carcasses. These data have been used for estimating the parameters in a prediction formula for prediction of lean meat percentage on the basis of the thickness measurements on the carcass. Data are described in detail in Busk et al. (1999). The first lines of data are

```
> library(gRbase)
> data(carcass)
> head(carcass)
```

S. Højsgaard et al., *Graphical Models with R*, Use R!,   77
DOI 10.1007/978-1-4614-2299-0_4, © Springer Science+Business Media, LLC 2012

	Fat11	Meat11	Fat12	Meat12	Fat13	Meat13	LeanMeat
1	17	51	12	51	12	61	56.52
2	17	49	15	48	15	54	57.58
3	14	38	11	34	11	40	55.89
4	17	58	12	58	11	58	61.82
5	14	51	12	48	13	54	62.96
6	20	40	14	40	14	45	54.58

Here Fat12 is the thickness of the fat layer between rib number 12 and 13 measured from the cranial part of the carcass, etc. LeanMeat is the percentage of meat in the carcass calculated as the weight of meat divided by the total weight of the carcass, consisting of meat, fat, and other tissues.

Gaussian graphical models provide a framework for modelling how these variables are mutually related. Consider a random vector $y = (y_1, \ldots, y_d)$ following a *multivariate normal* $\mathcal{N}_d(\mu, \Sigma)$ distribution. The key quantity in Gaussian graphical models is the inverse of the covariance matrix $K = \Sigma^{-1}$ known as the *concentration matrix*:

$$K = \begin{pmatrix} k_{11} & k_{12} & \cdots & k_{1d} \\ k_{21} & k_{22} & \cdots & k_{2d} \\ \vdots & \vdots & \ddots & \vdots \\ k_{d1} & k_{d2} & \cdots & k_{dd} \end{pmatrix}. \tag{4.1}$$

The *partial correlation* between y_u and y_v given all other variables can be simply derived from K as

$$\rho_{uv|V \setminus \{u,v\}} = -k_{uv}/\sqrt{k_{uu}k_{vv}}. \tag{4.2}$$

Thus $k_{uv} = 0$ if and only if y_u and y_v are conditionally independent given all other variables. In contrast to the concentrations, the partial correlations are invariant under a change of scale and origin in the sense that if $Y_v^* = a_v Y_v + b_v, v = 1, \ldots, d$ then $a_u a_v k_{uv}^* = k_{uv}$ and

$$\rho_{uv|V \setminus \{u,v\}}^* = \rho_{uv|V \setminus \{u,v\}}$$

where k^*, ρ^* are concentrations and partial correlations for Y^*.

Returning to the carcass data, the concentration matrix can be estimated as

```
> S.carc <- cov.wt (carcass, method="ML")$cov
> K.carc <-solve(S.carc)
> round(100*K.carc)
```

	Fat11	Meat11	Fat12	Meat12	Fat13	Meat13	LeanMeat
Fat11	44	3	-20	-7	-16	4	10
Meat11	3	16	-3	-6	-6	-6	-3
Fat12	-20	-3	54	6	-21	-5	9
Meat12	-7	-6	6	14	-1	-9	0
Fat13	-16	-6	-21	-1	56	3	7
Meat13	4	-6	-5	-9	3	16	-1
LeanMeat	10	-3	9	0	7	-1	26

Concentrations depend on the scale on which the variables are measured. Since lean meat percentage is measured on a different scale than the meat and fat measure-

Fig. 4.1 Model for the
`carcass` data found by
stepwise selection using the
AIC criterion

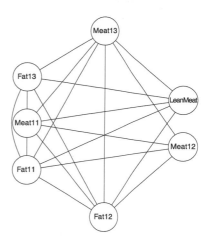

ments, the partial correlation matrix in (4.2)) seems more appropriate for measuring dependence. The `cov2pcor()` function provides such a measure:

```
> PC.carc <- cov2pcor(S.carc)
> round(100*PC.carc)
```

	Fat11	Meat11	Fat12	Meat12	Fat13	Meat13	LeanMeat
Fat11	100	-11	41	30	32	-16	-29
Meat11	-11	100	9	41	19	35	16
Fat12	41	9	100	-24	38	18	-24
Meat12	30	41	-24	100	2	61	2
Fat13	32	19	38	2	100	-9	-18
Meat13	-16	35	18	61	-9	100	7
LeanMeat	-29	16	-24	2	-18	7	100

Two of the partial correlations relating to `LeanMeat` are very small which suggests that `Fat13` is conditionally independent of `Meat12` given the remaining variables and this also holds for `LeanMeat` and `Meat12`. In particular this implies that `Meat12` can be left out without loss of accuracy in the linear prediction of `LeanMeat` from the meat and fat measurements.

A stepwise backward model selection procedure using AIC from the saturated model yields the model shown in Fig. 4.1. This has two edges removed, corresponding to the conditional independences observed above.

```
> sat.carc <- cmod(~.^., data=carcass)
> aic.carc<- stepwise(sat.carc)
> library(Rgraphviz)
> plot(as(aic.carc,"graphNEL"),"fdp")
```

Using BIC, yielding a higher penalty for complexity, we get a simpler graph, where also edges between `Fat13` and `Meat13` as well as `LeanMeat` and `Meat13` are removed:

```
> bic.carc<-stepwise(sat.carc,k=log(nrow(carcass)))
> bic.carc
```

```
Model: A cModel with 7 variables
 graphical : TRUE decomposable :  TRUE
```

Fig. 4.2 Model for the
`carcass` data found by
stepwise selection using the
BIC criterion

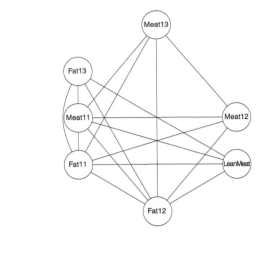

```
-2logL    :       11376.07 mdim :   24 aic :      11424.07
ideviance :        2465.16 idf  :   17 bic :      11516.25
deviance  :           8.62 df   :    4
> plot(as(bic.carc,"graphNEL"),"fdp")
```

This model is displayed in Fig. 4.2. It specifies a model with only two cliques. The
conditional independence relations can be summarised as

$$(\texttt{LeanMeat},\texttt{Fat13}) \perp\!\!\!\perp (\texttt{Meat12},\texttt{Meat13})|(\texttt{Fat11},\texttt{Fat12},\texttt{Meat11}).$$

In this model neither `Meat12` nor `Meat13` contribute directly to the prediction of
`LeanMeat`.

4.2.2 Body Fat Data

These data were collected by Dr. A. Garth Fisher and used by Johnson (1996) to illustrate issues in making multiple regression models for prediction of percentage of
body fat for an individual based on simple measurements of weight, circumferences
of body parts, etc. The data are available from StatLib and included in gRbase.

```
> data(BodyFat)
> head(BodyFat)

  Density BodyFat Age Weight Height Neck Chest Abdomen   Hip Thigh
1   1.071    12.3  23  154.2  67.75 36.2  93.1    85.2  94.5  59.0
2   1.085     6.1  22  173.2  72.25 38.5  93.6    83.0  98.7  58.7
3   1.041    25.3  22  154.0  66.25 34.0  95.8    87.9  99.2  59.6
4   1.075    10.4  26  184.8  72.25 37.4 101.8    86.4 101.2  60.1
5   1.034    28.7  24  184.2  71.25 34.4  97.3   100.0 101.9  63.2
6   1.050    20.9  24  210.2  74.75 39.0 104.5    94.4 107.8  66.0
  Knee Ankle Biceps Forearm Wrist
1 37.3  21.9   32.0    27.4  17.1
2 37.3  23.4   30.5    28.9  18.2
```

Fig. 4.3 Estimated
percentage of body fat
vs. reciprocal body density

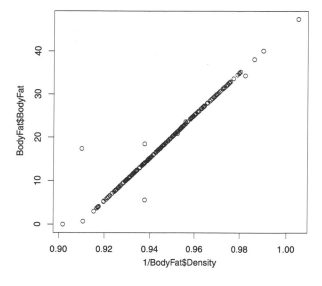

3	38.9	24.0	28.8	25.2	16.6
4	37.3	22.8	32.4	29.4	18.2
5	42.2	24.0	32.2	27.7	17.7
6	42.0	25.6	35.7	30.6	18.8

The measurement of body density is made by an elaborate underwater weighing procedure and the estimated percentage of body fat is then calculated from the body density by a linear expression in the reciprocal of the latter. However, as Fig. 4.3 shows, there are obvious errors in the data, as also indicated in Johnson (1996).

These obvious errors, and a few other similar oddities in the scatterplots, prompt us to reduce the data set by removing 11 strange observations. We also remove Density from the analysis as it is functionally related to BodyFat. In addition, we transform Age and Weight by the square root so we can better expect roughly linear relationships among the variables.

```
> BodyFat <- BodyFat[-c(31,42,48,76,86,96,159,169,175,182,206),]
> BodyFat$Age <- sqrt(BodyFat$Age)
> BodyFat$Weight <- sqrt(BodyFat$Weight)
> gRbodyfat <- BodyFat[,2:15]
```

After these changes, we get the following partial correlation matrix

```
> S.body<-cov.wt(gRbodyfat, method ="ML")$cov
> PC.body<-cov2pcor(S.body)
> round(100*PC.body)
```

	BodyFat	Age	Weight	Height	Neck	Chest	Abdomen	Hip	Thigh	Knee
BodyFat	100	11	2	-8	-17	-8	54	-16	8	0
Age	11	100	-16	-1	10	7	27	-9	-29	26
Weight	2	-16	100	73	29	54	33	46	18	10
Height	-8	-1	73	100	-17	-44	-27	-24	-27	17
Neck	-17	10	29	-17	100	-4	6	-16	9	-9
Chest	-8	7	54	-44	-4	100	25	-20	-24	-5
Abdomen	54	27	33	-27	6	25	100	22	-2	2

```
Hip          -16  -9    46  -24  -16  -20     22  100    32    5
Thigh          8 -29    18  -27    9  -24     -2   32   100   25
Knee           0  26    10   17   -9   -5      2    5    25  100
Ankle         -4 -18    25  -14  -15   -5    -13   -9     8   27
Biceps         2   7    16  -11    3    2    -11   -1    20   -3
Forearm       14 -17    19   -9   12    9    -23  -20     4    7
Wrist        -19  37     8   -1   23  -10      5    3   -13   -1
             Ankle Biceps Forearm Wrist
BodyFat       -4    2    14  -19
Age          -18    7   -17   37
Weight        25   16    19    8
Height       -14  -11    -9   -1
Neck         -15    3    12   23
Chest         -5    2     9  -10
Abdomen      -13  -11   -23    5
Hip           -9   -1   -20    3
Thigh          8   20     4  -13
Knee          27   -3     7   -1
Ankle        100  -13    -8   39
Biceps       -13  100    38    1
Forearm       -8   38   100   31
Wrist         39    1    31  100
```

where we in particular note the high partial correlation between BodyFat and Abdomen. If we again fit a model by stepwise selection using BIC we get

```
> sat.body <- cmod(~.^., data=gRbodyfat)
> bic.body<-stepwise(sat.body,k=log(nrow(gRbodyfat)))
> bic.body
```

```
Model: A cModel with 14 variables
 graphical :   TRUE  decomposable :  TRUE
 -2logL    :        12362.11 mdim :     75 aic :        12512.11
 ideviance :         4356.77 idf  :     61 bic :        12773.47
 deviance  :           35.67 df   :     30
```

```
> plot(bic.body,"neato")
```

and this model is displayed in Fig. 4.4. It has all variables except Chest, Knee, and Biceps as direct predictors for BodyFat. Note that the model is rather dense. Indeed 61 out of 91 possible edges are present in the model:

```
> graph::degree(as(bic.body,"graphNEL"))
```

Weight	Thigh	Height	Abdomen	Forearm	BodyFat	Neck	Hip
12	12	11	9	11	10	10	8
Age	Chest	Knee	Ankle	Biceps	Wrist		
10	5	5	10	4	5		

Note that we need to use graph::degree to avoid conflicts between the **graph** and **igraph** packages which both feature the degree function.

4.3 Undirected Gaussian Graphical Models

In this section we consider models for Gaussian data that can be represented as undirected graphs, as illustrated in the previous section.

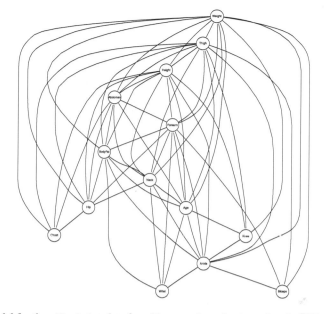

Fig. 4.4 Model for the `gRbodyfat` data found by stepwise selection using the BIC criterion

4.3.1 Preliminaries and Notation

An undirected *Gaussian graphical model* (hereafter abbreviated *UGGM*) is represented by an undirected graph $\mathcal{G} = (\Gamma, E)$ where the vertices $\Gamma = \{1, \ldots, d\}$ represent the set of variables and E is a set of undirected edges.

When a random vector y follows a Gaussian distribution $\mathcal{N}_d(\mu, \Sigma)$, the graph $\mathcal{G}$ represents the model where $K = \Sigma^{-1}$ is a positive definite matrix with $k_{uv} = 0$ whenever there is no edge between vertices u and v in $\mathcal{G}$. This graph is called the *dependence graph* of the model because it holds for all u, v that if u and v are not adjacent, it holds that $u \perp\!\!\!\perp v | \Gamma \setminus \{u, v\}$.

It is often convenient to represent the model by the cliques $\mathcal{C} = \{C_1, \ldots, C_Q\}$ of the dependence graph. Recall that a probability distribution factorizes according to an undirected graph $\mathcal{G} = (V, E)$ if it admits a factorisation of the form

$$f(x_V) = \prod_{i=1\ldots k} g_i(x_{C_i})$$

where $C_1 \ldots C_k$ are the cliques of $\mathcal{G}$, and $g_1() \ldots g_k()$ are arbitrary functions. In analogy with the terminology for log-linear models in Chap. 2 we also say that $\mathcal{C}$ is the *generating class* for the model.

As for log-linear models, the generating class can be specified by a model formula or by a list. For example the model `gen.carc` below for the `carcass data` specifies the UGGM with edges missing for all partial correlations less than or equal to .12.

Fig. 4.5 Model for the
carcass data found by
thresholding partial
correlations

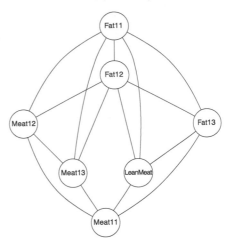

```
> gen.carc<-cmod(~Fat11 * Fat12 * Meat12 * Meat13 +
+                 Fat11 * Fat12 * Fat13 * LeanMeat +
+                 Meat11 * Meat12 * Meat13 +
+                 Meat11 * Fat13 * LeanMeat,data=carcass)
> gen.carc
```

```
Model: A cModel with 7 variables
 graphical :  TRUE  decomposable : FALSE
 -2logL    :      11387.24 mdim :   22 aic :      11431.24
 ideviance :       2453.99 idf  :   15 bic :      11515.73
 deviance  :         19.79 df   :    6
```

```
> plot(gen.carc,"neato")
```

This model is shown in Fig. 4.5.

Recall that $K = \Sigma^{-1}$ and define $h = K\mu$. The multivariate normal density is
then

$$f(y) = (2\pi)^{-d/2}\det(K)^{\frac{1}{2}}\exp\left\{-\frac{1}{2}(y-\mu)^{\top}K(y-\mu)\right\}$$

$$= (2\pi)^{-d/2}\det(K)^{\frac{1}{2}}\exp\left(-\frac{1}{2}\mu^{\top}K\mu + h^{\top}y - \frac{1}{2}y^{\top}Ky\right).$$

Letting $a = -\frac{d}{2}\log(2\pi) + \frac{1}{2}\log\det(K) - \frac{1}{2}\mu^{\top}K\mu$ we get

$$f(y) = \exp\left(a + h^{\top}y - \frac{1}{2}y^{\top}Ky\right)$$

$$= \exp\left(a + \sum_{u}h_{u}y_{u} - \frac{1}{2}\sum_{u}\sum_{v}k_{uv}y_{u}y_{v}\right). \tag{4.3}$$

If the sets of vertices A and B are separated by a set C in the dependence
graph we have $k_{uv} = 0$ for $u \in A$ and $v \in B$. By appropriately collecting terms we

can write $f(y) = g(y_A, y_C)h(y_B, y_C)$. The factorization criterion applied to (4.3) now yields that for Gaussian graphical models the *global Markov property* holds: $A \perp\!\!\!\perp B \,|\, C$.

Moreover, note that there are at most pairwise interactions between variables. That is, in contrast to log-linear models, there is no such thing as third and higher order interactions for Gaussian graphical models. Hence the model is completely determined by the edges of the dependence graph. For example we could alternatively specify gen.carc above as

```
> edge.carc<-cmod(edgeList(as(gen.carc,"graphNEL")),data=carcass)
> edge.carc
```

```
Model: A cModel with 7 variables
 graphical :   TRUE  decomposable : FALSE
 -2logL    :       11387.24 mdim :    22 aic :    11431.24
 ideviance :        2453.99 idf  :    15 bic :    11515.73
 deviance  :          19.79 df   :     6
```

where we have only specified pairwise interactions.

4.3.2 Estimation, Likelihood, and Model Fitting

Consider a sample $y^1, \ldots, y^n$ of n observations of $y \sim \mathcal{N}_d(\mu, \Sigma)$. Let W denote the matrix of sums of squares and products, $W = \sum_{\nu=1}^{n}(y^\nu - \bar{y})(y^\nu - \bar{y})^\top$, and let $S = W/n$ denote the empirical covariance matrix. The log-likelihood function based on the sample is

$$\log L(K, \mu) = \frac{n}{2} \log \det(K) - \frac{n}{2} \mathrm{tr}(KS) - \frac{n}{2}(\bar{y} - \mu)^\top K(\bar{y} - \mu). \qquad (4.4)$$

For fixed K this is clearly maximized for $\hat{\mu} = \bar{x}$ which renders the last term equal to zero. The profile likelihood for K thus becomes

$$\log L(K, \hat{\mu}) = \frac{n}{2} \log \det(K) - \frac{n}{2} \mathrm{tr}(KS). \qquad (4.5)$$

Since $\mathrm{tr}(KS) = \sum_u \sum_v s_{uv} k_{uv}$ it follows that only elements s_{uv} of S for which the corresponding elements k_{uv} of K are non-zero will contribute to the likelihood.

For a matrix M with entries m_{uv} for $u \in \Gamma$ and $v \in \Gamma$ and two subsets $A \subset \Gamma$ and $B \subset \Gamma$ we let M_{AB} denote the corresponding $|A| \times |B|$ submatrix of M with entries m_{uv} for $u \in A$ and $v \in B$.

If the UGGM has generating class $\mathcal{C} = \{C_1, \ldots, C_Q\}$ it can be seen that the submatrices $S_{C_l C_l}$, for $l = 1, \ldots, Q$ together with the sample mean $\bar{y}$ jointly form a set of minimal sufficient statistics. The maximum likelihood estimate is determined as the unique solution to the system of equations

$$\hat{\mu} = \bar{x}, \qquad \hat{\Sigma}_{C_l C_l} = S_{C_l C_l}, \quad l = 1, \ldots, Q \qquad (4.6)$$

which satisfies the restrictions on the concentration matrix. Thus the MLE of the covariance between any pair of variables which are neighbours in the graph is equal to the corresponding empirical quantity. It follows that the MLE for Σ under the saturated model with no conditional independence restrictions satisfies $\hat{\Sigma} = S$ so in that case we have $\hat{K} = S^{-1}$, provided S is not singular.

In general (4.6) must be solved iteratively, for example using the IPS (iterative proportional scaling) algorithm (Dempster 1972; Speed and Kiiveri 1986). However, Sect. 4.3.5 describes models for which estimation can be made in closed form.

The starting point for K for the IPS algorithm can be any positive definite matrix satisfying the constraints of the model; for example the identity matrix. Let $C \in \mathcal{C}$ be one of the generators and let $B = \Gamma \setminus C$. The submatrix K_{CC} of K is modified by an increment E so as to satisfy the constraints by the likelihood equations, that is $\hat{\Sigma}_{CC} = S_{CC}$. The increment can be found as follows: The requirement from the likelihood equations is that

$$\begin{pmatrix} S_{CC} & S_{CB} \\ S_{BC} & S_{BB} \end{pmatrix} = \begin{pmatrix} K_{CC} + E & K_{CB} \\ K_{BC} & K_{BB} \end{pmatrix}^{-1}.$$

Standard results on the inverse of partitioned matrices gives

$$S_{CC} = (K_{CC} + E - K_{CB} K_{BB}^{-1} K_{BC})^{-1}.$$

Here and elsewhere expressions such as M_{CC}^{-1} should be read as the inverse of the submatrix M_{CC}, i.e. $M_{CC}^{-1} = (M_{CC})^{-1}$. Hence we have

$$E = S_{CC}^{-1} - (K_{CC} - K_{CB} K_{BB}^{-1} K_{BC}) = S_{CC}^{-1} - \Sigma_{CC}^{-1}$$

so K_{CC} is to be replaced with $K_{CC} + S_{CC}^{-1} - \Sigma_{CC}^{-1}$, that is

$$K_{CC} \leftarrow S_{CC}^{-1} + K_{CB} K_{BB}^{-1} K_{BC}. \tag{4.7}$$

Each step of (4.7) will lead to an non-decreasing likelihood and iterative proportional scaling consists in repeatedly cycling through the generators $C_1, \ldots, C_Q$ and updating K as above until convergence. The IPS algorithm is implemented in the function ggmfit():

```
> carcfit1 <-
+ ggmfit(S.carc,n=nrow(carcass),edgeList(as(gen.carc,"graphNEL")))
> carcfit1[c("dev","df","iter")]

$dev
[1] 19.79

$df
[1] 6

$iter
[1] 774
```

The deviance, degrees of freedom and number of iterations required are shown. The object `carcfit1` contains much additional information, including the parameter estimates. Note that specifying the generators as edges via an `edgeList` is relatively inefficient: it is much more efficient to specify the cliques of the graph. Each step of the algorithm is computationally simpler, and since the parameters move in blocks, the algorithm converges faster.

```
> cgens <- maxClique(as(gen.carc,"graphNEL"))$maxCliques
> carcfit2 <- ggmfit(S.carc, n=nrow(carcass), cgens)
> carcfit2[c("dev","df","iter")]

$dev
[1] 19.79

$df
[1] 6

$iter
[1] 61
```

4.3.3 Hypothesis Testing

The maximized value of the likelihood can be found as follows from (4.5): Because $\hat{\Sigma}$ and S differ exactly on those entries for which $k_{uv} = 0$ it holds $\text{tr}(\hat{K}S) = \text{tr}(\hat{K}\hat{\Sigma}) = d$. Hence the maximized value of the log likelihood is $\hat{l} = n \log \det(\hat{K})/2 - nd/2$. Thus the *deviance* of a model $\mathcal{M}$ is

$$D = \text{dev} = 2(\hat{l}_{\text{sat}} - \hat{l}) = n \log\{\det(S^{-1})/\det(\hat{K})\} = -n \log \det(S\hat{K}) \qquad (4.8)$$

and similarly the *ideviance* representing the log-likelihood ratio relative to the independence model is

$$iD = \text{idev} = 2(\hat{l} - \hat{l}_{\text{ind}}) = n \left\{ \log \det(\hat{K}) + \sum_{i=1}^{d} \log s_{ii} \right\} \qquad (4.9)$$

which makes sense even if the saturated model cannot be fitted. The likelihood ratio test statistic for testing $\mathcal{M}_1$ under $\mathcal{M}_0$ where $\mathcal{M}_1 \subseteq \mathcal{M}_0$ is the difference in deviance (or ideviance) between the two models:

$$\text{lrt} = 2(\hat{l}_0 - \hat{l}_1) = n \log(\det(\hat{K}_0)/\det(\hat{K}_1))$$

The likelihood ratio test statistic can be used for testing $\mathcal{M}_1$ under $\mathcal{M}_0$: large values of `lrt` suggest that the null hypothesis $\mathcal{M}_1$ is false. Under the hypothesis that $\mathcal{M}_1$ holds, `lrt` has an approximate χ^2_f distribution where f is the difference in the number of parameters of the two models, which is the same as the difference in the number of edges.

```
> comparemodels <- function(m1,m2) {
+    lrt <- m2$fitinfo$dev - m1$fitinfo$dev
+    dfdiff <- m2$fitinfo$dimension[4] - m1$fitinfo$dimension[4]
+    names(dfdiff) <- NULL
+    list('lrt'=lrt, 'df'=dfdiff)
+ }
> comparemodels(aic.carc,bic.carc)

$lrt
[1] 8.373

$df
[1] 2
```

indicating that the simpler model does not quite fit.

The ciTest_mvn() function can be used for testing a single conditional independence hypothesis, or when put in the terminology of graphs, for testing whether
a single edge can be deleted from the saturated model (the model for which the
dependence graph is complete). Default is to use the likelihood ratio test for testing against the saturated model which is the deviance (4.8). For example to test for
LeanMeat $\perp\!\!\!\perp$ Meat13|rest we can use:

```
> ciTest_mvn(list(cov=S.carc, n.obs=nrow(carcass)),
+            set = ~LeanMeat+Meat13+Meat11+Meat12+Fat11+Fat12+Fat13)

Testing LeanMeat _|_ Meat13 | Meat11 Meat12 Fat11 Fat12 Fat13
Statistic (DEV):    1.687 df: 1 p-value: 0.1940 method: CHISQ
```

so ciTest_mvn() interprets set by testing conditional independence of the two
first variables given the remaining.

Alternative test statistics exist which are more accurate for small samples. An
option for ciTest_mvn() is the F statistic

$$F = (n - d)(e^{\text{dev}/n} - 1) \tag{4.10}$$

which has an $F_{1,n-d}$ distribution under the hypothesis, i.e. $\sqrt{F}$ is distributed as the
absolute value of a Student's t.

```
> ciTest_mvn(list(cov=S.carc, n.obs=nrow(carcass)),
+            set=~LeanMeat+Meat11+Meat12+Meat13+Fat11+Fat12+Fat13,
+            statistic="F")

Testing LeanMeat _|_ Meat11 | Meat12 Meat13 Fat11 Fat12 Fat13
Statistic (F):    8.864 df: 1 p-value: 0.0031 method: F
```

Another possibility is to base the test on asymptotic normality of Fisher's z transform of the partial correlation; this is done using the gaussCItest() in the **pcalg**
package:

```
> library(pcalg)
> C.carc<-cov2cor(S.carc)
> gaussCItest(7,2,c(1,3,4,5,6),list(C=C.carc,n=nrow(carcass)))

[1] 0.003077
```

Note that in this case, with $n - d = 339$ we find almost exactly the same p-value as
the F-test. In fact, the p-value for the F statistic is 0.003118 with four significant
digits.

4.3.4 Concentration and Regression

There is a close connection between the concentration matrix K in (4.1) and multiple linear regression. Let $A \subset \Gamma$ and $B = \Gamma \setminus A$. This defines a partitioning of y, μ, Σ and K according to A and B as

$$y = \begin{pmatrix} y_A \\ y_B \end{pmatrix}, \qquad \mu = \begin{pmatrix} \mu_A \\ \mu_B \end{pmatrix},$$

$$\Sigma = \begin{pmatrix} \Sigma_{AA} & \Sigma_{AB} \\ \Sigma_{BA} & \Sigma_{BB} \end{pmatrix}, \qquad K = \begin{pmatrix} K_{AA} & K_{AB} \\ K_{BA} & K_{BB} \end{pmatrix}.$$

Then

$$y_A | y_B \sim \mathcal{N}(\mu_{A|B}, \Sigma_{A|B})$$

where

$$\mu_{A|B} = \mu_A + \Sigma_{AB} \Sigma_{BB}^{-1} (y_B - \mu_B), \tag{4.11}$$

$$\Sigma_{A|B} = \Sigma_{AA} - \Sigma_{AB} \Sigma_{BB}^{-1} \Sigma_{AB}. \tag{4.12}$$

Standard results on the inverse of partitioned matrices gives that the quantities in (4.11) can be expressed in terms of K as

$$\Sigma_{AB} \Sigma_{BB}^{-1} = -K_{AA}^{-1} K_{AB} \quad \text{and} \quad \Sigma_{AA} - \Sigma_{AB} \Sigma_{BB}^{-1} \Sigma_{AB} = K_{AA}^{-1}.$$

To illustrate this, consider a multiple regression of y_1 on $y_2, \ldots, y_d$ as explanatory variables,

$$y_1 = a_1 + \beta_{13} y_2 + \cdots + \beta_{1d} y_d + \epsilon_1 \quad \text{where } \epsilon_1 \sim \mathcal{N}(0, \sigma_1^2). \tag{4.13}$$

Then $\sigma_1^2 = 1/k_{11}$ while the regression coefficients $\beta_{12}, \ldots, \beta_{1d}$ can be derived from K as

$$(\beta_{12}, \ldots, \beta_{1d}) = -(k_{12}, \ldots, k_{1d})/k_{11}.$$

Returning to K for the carcass data we find that the regression coefficients for predicting LeanMeat are

```
> -K.carc[7,-7]/K.carc[7,7]
```

```
   Fat11    Meat11    Fat12    Meat12    Fat13    Meat13
-0.37715   0.12422  -0.33552   0.01518  -0.26893  0.05414
```

while the residual variance of the lean meat percentage is

```
> 1/K.carc[7,7]
```

```
[1] 3.794
```

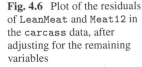

Fig. 4.6 Plot of the residuals of `LeanMeat` and `Meat12` in the `carcass` data, after adjusting for the remaining variables

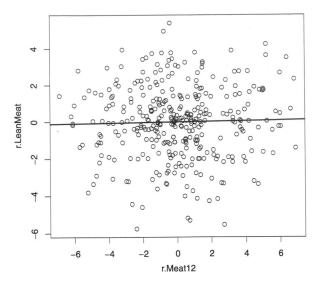

Notice that the regression coefficients to `Meat12` and `Meat13` are very small, corresponding to the conditional independence structure found in model `bic.carc`, implying these are redundant for prediction of `LeanMeat`. The matrix

$$\begin{pmatrix} k_{11} & k_{12} \\ k_{21} & k_{22} \end{pmatrix}^{-1} = \frac{1}{k_{11}k_{22} - k_{12}^2} \begin{pmatrix} k_{22} & -k_{12} \\ -k_{21} & k_{11} \end{pmatrix} \qquad (4.14)$$

is the conditional covariance matrix for (y_1, y_2) given $y_3, \ldots y_d$. That is, the covariance between the residuals of y_1 and y_2 after adjusting for the effect of all other variables. This observation provides a diagnostic tool for investigating conditional independence: The residuals after regression should be uncorrelated. It is apparent from Fig. 4.6 that this is entirely reasonable.

```
> r.LeanMeat <- residuals(lm(LeanMeat~Meat11+Meat13+Fat11+Fat12+Fat13,
+   data=carcass))
> r.Meat12 <- residuals(lm(Meat12~Meat11+Meat13+Fat11+Fat12+Fat13,
+   data=carcass))
> plot(r.Meat12,r.LeanMeat )
> abline(lm(r.LeanMeat~r.Meat12),lwd=2)
```

4.3.5 Decomposition of UGGMs

As mentioned in Sect. 4.3.2, some models have the property that estimation can be made in closed form, i.e. without iteration.

To describe these we introduce the following notation: for a given $|A| \times |B|$ matrix M_{AB} we let $[M_{AB}]^{\Gamma}$ denote the $|\Gamma| \times |\Gamma|$ matrix obtained from M_{AB} by filling up with zeros to obtain the full dimension.

Let $\mathcal{M}$ be a UGGM with variables Γ and let $\mathcal{G}$ be the dependence graph for $\mathcal{M}$. For a $A \subset \Gamma$, the A-marginal model $\mathcal{M}_A$ is the UGGM induced by the subgraph $\mathcal{G}_A$ of $\mathcal{G}$.

Suppose that (A, B, C) is a decomposition of a dependence graph of $\mathcal{M}$ such that (i) A, B and C are disjoint, (ii) $A \cup B \cup C = \Gamma$, (iii) all paths from a variable in A to a variable in B goes through C and (iv) C is complete in $\mathcal{G}$, i.e. C is a complete separator of $\mathcal{G}$. Then the MLE of K can be found as

$$\hat{K} = [\hat{K}_{A \cup C}]^\Gamma + [\hat{K}_{B \cup C}]^\Gamma - [S_C^{-1}]^\Gamma \qquad (4.15)$$

where $\hat{K}_{A \cup C}$ is the MLE for $K_{A \cup C}$ under the marginal model $\mathcal{M}_{A \cup C}$ and likewise for $\hat{K}_{B \cup C}$. The determinant of $\hat{K}$ appears in the expression for the deviance in (4.8) and this determinant can be factorized in a way similar to (4.15),

$$\det(\hat{K}) = \det(\hat{K}_{A \cup C}) \det(\hat{K}_{B \cup C}) / \det(\hat{S}_C^{-1}). \qquad (4.16)$$

As an illustration, consider the model displayed in Fig. 4.2 and let $A = \{\texttt{Fat13}, \texttt{LeanMeat}\}$, $B = \{\texttt{Meat12}, \texttt{Meat13}\}$ and $C = \{\texttt{Fat11}, \texttt{Fat12}, \texttt{Meat11}\}$. Then the triple (A, B, C) is a decomposition of its dependence graph. Secondly, the marginal models $\mathcal{M}_{A \cup C}$ and $\mathcal{M}_{B \cup C}$ are both saturated models so for these $\hat{K}_{A \cup C} = S_{A \cup C}^{-1}$ and $\hat{K}_{B \cup C} = S_{B \cup C}^{-1}$. Therefore the MLE of K can be found using (4.15) as follows:

```
> K.hat<-S.carc
> K.hat[]<-0
> AC <- c("Fat11","Fat12","Fat13","Meat11","LeanMeat")
> BC <- c("Meat11","Meat12","Meat13","Fat11","Fat12")
> C <- c("Fat11","Fat12","Meat11")
> K.hat[AC,AC] <- K.hat[AC,AC] + solve(S.carc[AC,AC])
> K.hat[BC,BC] <- K.hat[BC,BC] + solve(S.carc[BC,BC])
> K.hat[C,C] <- K.hat[C,C] - solve(S.carc[C,C])
> round(100*K.hat)
```

	Fat11	Meat11	Fat12	Meat12	Fat13	Meat13	LeanMeat
Fat11	44	1	-20	-7	-16	6	10
Meat11	1	16	-4	-6	-4	-5	-5
Fat12	-20	-4	54	6	-20	-4	9
Meat12	-7	-6	6	14	0	-9	0
Fat13	-16	-4	-20	0	55	0	7
Meat13	6	-5	-4	-9	0	16	0
LeanMeat	10	-5	9	0	7	0	26

```
> Sigma.hat <- solve(K.hat)
> round(Sigma.hat,2)
```

	Fat11	Meat11	Fat12	Meat12	Fat13	Meat13	LeanMeat
Fat11	11.34	0.74	8.42	2.06	7.66	-0.76	-9.08
Meat11	0.74	32.97	0.67	35.94	2.01	31.97	5.33
Fat12	8.42	0.67	8.91	0.31	6.84	-0.60	-7.95
Meat12	2.06	35.94	0.31	51.79	2.45	41.47	5.41
Fat13	7.66	2.01	6.84	2.45	7.62	0.89	-6.93
Meat13	-0.76	31.97	-0.60	41.47	0.89	41.44	6.43
LeanMeat	-9.08	5.33	-7.95	5.41	-6.93	6.43	12.90

```
> round(S.carc,2)

          Fat11 Meat11 Fat12 Meat12 Fat13 Meat13 LeanMeat
Fat11     11.34   0.74  8.42   2.06  7.66  -0.76    -9.08
Meat11     0.74  32.97  0.67  35.94  2.01  31.97     5.33
Fat12      8.42   0.67  8.91   0.31  6.84  -0.60    -7.95
Meat12     2.06  35.94  0.31  51.79  2.18  41.47     6.03
Fat13      7.66   2.01  6.84   2.18  7.62   0.38    -6.93
Meat13    -0.76  31.97 -0.60  41.47  0.38  41.44     7.23
LeanMeat  -9.08   5.33 -7.95   6.03 -6.93   7.23    12.90
```

This example also demonstrates the likelihood equations in (4.6): S.carc and Sigma.hat match on entries which correspond to edges in the dependence graph

An UGGM is *decomposable* if its dependence graph is triangulated. For a decomposable UGGM, the MLE for K can be found by successively applying (4.15); see Lauritzen (1996), p. 145 for an explicit expression.

4.4 Model Selection

For a general description of model selection issues for graphical models, please see Chap. 2. For UGGMs the problems are typically simpler, as one can concentrate on identifying the graph rather than more general forms of model generating classes.

A practical problem with stepwise selection strategies is that they tend to become time consuming for problems with many variables and usually only a small part of the relevant search space is covered during a search.

For UGGMs specific types of alternative model identification methods have been developed in this case exploiting the simplicity of UGGMs when compared to log-linear models. We describe and illustrate some of these on the carcass data example. See also Chap. 7 for methods that are specifically tailored to high-dimensional cases.

4.4.1 Stepwise Methods

The stepwise() function in **gRim** performs stepwise model selection, based on a variety of criteria. Use of the AIC and BIC was illustrated above in Sect. 4.2.1. Alternatively, significance tests may be used, as we now illustrate. The default significance level is 0.05.

```
> test.carc <- stepwise(sat.carc, details=1,"test")

STEPWISE:
 criterion: test
 direction: backward
 type     : decomposable
 search   : all
 steps    : 1000
. BACKWARD: type=decomposable search=all, criterion=test, alpha=0.05
```

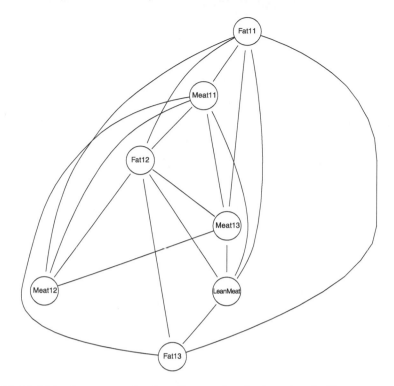

Fig. 4.7 Model for the `carcass` data found by stepwise selection using significance tests

```
. Initial model: is graphical=TRUE is decomposable=TRUE
  p.value    0.7089 Edge deleted: Fat13 Meat12
  p.value    0.7414 Edge deleted: Meat12 LeanMeat
  p.value    0.0812 Edge deleted: Fat13 Meat13

> plot(test.carc,"neato")
```

As the output shows, the default method is backward elimination of edges (as opposed to forward addition of edges). This results in the model shown in Fig. 4.7. It has one edge less than the model `aic.carc`; the edge between `Meat13` and `Fat13` is also removed.

The default selection criterion is the *Akaike information criterion* or *AIC* (Akaike 1974) which mimimizes the negative of a *penalized likelihood*

$$\text{AIC}(k) = -2\log L + k\dim(\mathcal{M})$$

where $\dim(\mathcal{M})$ is the number of independent parameters in model $\mathcal{M}$ and $k = 2$ for the AIC. Other values of k can be specified by the user as an option to `stepwise()`. For example, choosing $k = \log n$ where n is the number of observations gives the *Bayesian information criterion* or *BIC* (Schwarz 1978) as used for the model `bic.carc` below.

Default is to search among decomposable models (if the initial model is decomposable; otherwise the search will be among all models). The reason is that searching among decomposable models (as opposed to searching among general models) is faster: Suppose $\mathcal{M}_0$ is a decomposable model with cliques $\mathcal{C} = \{C_1, \ldots, C_Q\}$ and let $\hat{K}^0$ be the estimated concentration matrix under that model. Consider a model reduction where $\mathcal{M}_1$ is the model obtained by deleting an edge $e = \{u, v\}$ which belongs to only one clique $C \in \mathcal{C}$. Then $\mathcal{M}_1$ is also decomposable. The likelihood ratio test for $\mathcal{M}_1$ under $\mathcal{M}_0$ can be carried out as a test for $u \perp\!\!\!\perp v | C \setminus \{u, v\}$, that is a test for deletion of e from the saturated marginal model given by C. Thus test for deletion of e does not require a large model to be fitted. The reason is that the contributions to from all the other cliques than C to the likelihood are the same under the two models and hence cancel out in the likelihood ratio. Moreover, fitting $\mathcal{M}_1$ is straightforward if $\hat{K}^0$ is given: the triple $(\{u\}, \{v\}, C \setminus \{u, v\})$ is a decomposition of the C-marginal model (after e has been deleted). Hence $\hat{K}_C$ can be found from (4.15) and the submatrix $\hat{K}^0_{CC}$ of $\hat{K}^0$ is then to be replaced by $\hat{K}_C$ to obtain $\hat{K}^1$. A final computational saving is obtained by exploiting that several of the test statistics can be reused during a sequence of reductions.

Also forward selection among decomposable models is faster than searching among general models. If adding an edge e to a decomposable model $\mathcal{M}_0$ gives a model $\mathcal{M}_1$ which is also decomposable then the testing scheme above can also be applied. However, determining from $\mathcal{M}_0$ and e whether $\mathcal{M}_1$ is decomposable is not straightforward. The approach taken in cmod() is to actually create $\mathcal{M}_1$ and then check whether $\mathcal{M}_1$ is decomposable. This can be done using maximum cardinality search, see Sect. 1.4.1.

The default search method in **gRim** is to consider all edges in $\mathcal{M}_0$ and then choose the optimal one (according to the selection criterion) to delete. However, setting search ="headlong" will cause edges to be searched in random order. When an edge which can be deleted (or added) is found, this edge is deleted (or added) and the new model is formed. Thereby the headlong strategy can make model search considerably faster.

```
> ind.carc<-cmod(~.^1,data=carcass)
> set.seed(123)
> forw.carc<-stepwise(ind.carc,search="headlong",
+ direction="forward",k=log(nrow(carcass)),details=0)
> forw.carc

Model: A cModel with 7 variables
 graphical :   TRUE   decomposable :   TRUE
 -2logL    :        11393.53 mdim :    23 aic :     11439.53
 ideviance :         2447.70 idf  :    16 bic :     11527.87
 deviance  :           26.08 df   :     5

> plot(forw.carc,"neato")
```

This model, shown in Fig. 4.8, is considerably smaller than any of those previously found, although the only variable which can be ignored for prediction of LeanMeat is Meat11.

Fig. 4.8 Model for the carcass data found by headlong stepwise selection using the AIC criterion

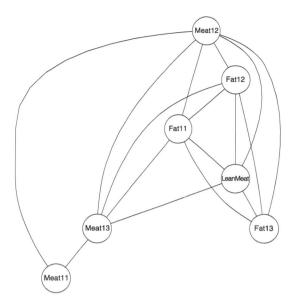

4.4.2 Convex Optimization

One way to avoid a stepwise search is to use the glasso algorithm (Friedman et al. 2008). This gives a fast technique to find the Gaussian graphical model that maximizes a log-likelihood for K which is penalized by the L_1-norm $|K|$; this is implemented in the package **glasso**. From (4.5) an L_1-penalized log-likelihood is equivalent to

$$L_{\text{pen}}(K, \hat{\mu}) = \log \det(K) - \text{tr}(KS) - \rho|K| \qquad (4.17)$$

where ρ is a non-negative penalty parameter. Thee L_1-norm $|K|$ is the sum of the absolute values of the elements of the concentration matrix K. This sum is largely a proxy for the number of non-zero elements of K and as this penalized log-likelihood is convex in K, it can be optimized by convex programming methods.

The smaller the value of ρ, the denser the graph that results. No penalization occurs for values of ρ close to zero. We illustrate the methods on the gRbodyfat data:

```
> C.body<-cov2cor(S.body)
> library(glasso)
> res.lasso<-glasso(C.body,rho=0.1)
> AM <- res.lasso$wi != 0
> diag(AM) <- F
> g.lasso <- as(AM, "graphNEL")
> nodes(g.lasso)<-names(gRbodyfat)
> glasso.body<-cmod(g.lasso,data=gRbodyfat)
> plot(glasso.body, "neato")
```

This graph, shown in Fig. 4.9, has a similar density with a total of 60 edges, but BodyFat is now only connected to Age, Height, Chest, Abdomen, Hip, and Thigh.

Fig. 4.9 Model for the
gRbodyfat data selected by
the glasso algorithm

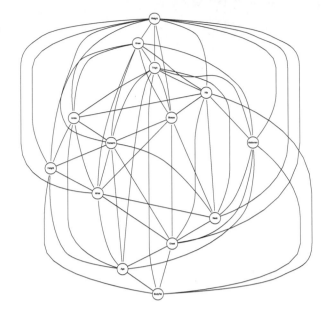

```
> graph::degree(as(glasso.body,"graphNEL"))
```

Weight	Knee	Thigh	Hip	Abdomen	Ankle	Biceps	Forearm
11	9	10	9	8	8	8	10

Height	Wrist	Neck	Chest	Age	BodyFat
7	9	8	9	8	6

4.4.3 Thresholding

A simple and apparently naive method for selecting a UGGM is to set a specific
threshold for the partial correlations, so edges are removed for all partial correlations
less than a given value.

```
> round(100*PC.carc)
```

	Fat11	Meat11	Fat12	Meat12	Fat13	Meat13	LeanMeat
Fat11	100	-11	41	30	32	-16	-29
Meat11	-11	100	9	41	19	35	16
Fat12	41	9	100	-24	38	18	-24
Meat12	30	41	-24	100	2	61	2
Fat13	32	19	38	2	100	-9	-18
Meat13	-16	35	18	61	-9	100	7
LeanMeat	-29	16	-24	2	-18	7	100

For example, we may stipulate that entries which are numerically smaller than e.g.
0.1 can be set to zero while the remaining values are set to one. Setting also the
diagonal elements to zero then yields an adjacency matrix:

Fig. 4.10 Model for the carcass data found by thresholding the partial correlations

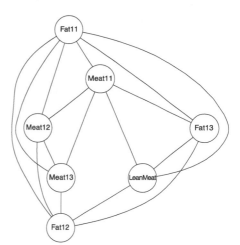

```
> threshold <- .1
> Z <- abs(PC.carc)
> Z[Z<threshold] <- 0
> diag(Z)<-0
> Z[Z>0] <- 1
> g.thresh<-as(Z, "graphNEL")
> thresh.carc <- cmod(g.thresh, data=carcass)
> thresh.carc
```

```
Model: A cModel with 7 variables
 graphical :   TRUE  decomposable : FALSE
 -2logL    :       11384.93 mdim :    23 aic :      11430.93
 ideviance :        2456.30 idf  :    16 bic :      11519.27
 deviance  :          17.48 df   :     5
```

```
> plot(thresh.carc,"neato")
```

The resulting model is shown in Fig. 4.10. Here also the edge between LeanMeat and Meat13 has been removed.

4.4.4 Simultaneous p-Values

A special form of thresholding is the approach implemented in the package **SIN**, due to Drton and Perlman (2007, 2008). Consider the set of hypotheses

$$\mathcal{H} = \{H_{uv} : Y_u \perp\!\!\!\perp Y_v | Y_{V \setminus \{u,v\}}\}_{u<v},$$

and let $\mathcal{P} = \{p_{uv}\}_{u<v}$ be the corresponding nominal p-values. Using Fisher's z-transform and an inequality of Sidak (1967) these are transformed to a set of *simultaneous* p-values $\tilde{\mathcal{P}} = \{\tilde{p}_{uv}\}$ which control the familywise error rate: that is to say, if we reject H_{uv} whenever $\tilde{p}_{uv} < \alpha$, the probability of rejecting one or more *true* hypotheses H_{uv} is less or equal to α. A sequential multiple testing procedure

Fig. 4.11 A plot of the
simultaneous *p*-values for the
carcass data

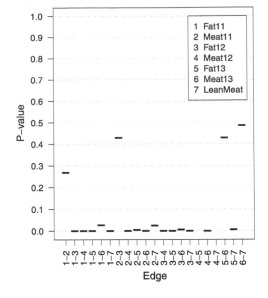

due to Holm (1979) can then further be applied to $\tilde{\mathcal{P}}$ to derive a modified set $\check{\mathcal{P}}$ of
simultaneous *p*-values with higher probability of rejecting H_{uv} when H_{uv} is in fact
false.

It follows that if we construct a graph $G(\alpha)$ by including precisely those edges
with $\check{p}_{uv} < \alpha$, then the probability of incorrectly including one or more edges is less
or equal to α. Or to put this another way, the probability that $G(\alpha)$ is not a subgraph
of the true model is less or equal to α.

The authors suggest using two α thresholds so as to partition the simultaneous *p*-
values into three sets: a significant set S, an intermediate set I and a non-significant
set N (hence the spicy acronym, SIN). The following code displays a plot of the
simultaneous *p*-values, see Fig. 4.11.

```
> library(SIN)
> psin.carc<-sinUG(S.carc,n=nrow(carcass))
> plotUGpvalues(psin.carc)
```

We may take $\alpha = 0.1$ and 0.3, for example.

```
> gsin.carc <- as(getgraph(psin.carc, 0.1), "graphNEL")
> plot(gsin.carc, "neato")
```

Figure 4.12 shows $G(0.1)$. Note that LeanMeat depends primarily on the fat variables, plus Meat11. Using $\alpha = 0.3$ would just add the edge between Fat11 and
Meat11, so we omit this here.

Other functions in the **SIN** package implement the same approach with different classes of models for Gaussian data, for example, DAGs with known variable
ordering, and chain graphs with known block structure.

We can perform a similar analysis of gRbodyfat:

```
> psin.body<-sinUG(S.body,n=nrow(gRbodyfat))
> plotUGpvalues(psin.body)
```

Fig. 4.12 Model for the `carcass` data found by thresholding the simultaneous *p*-values

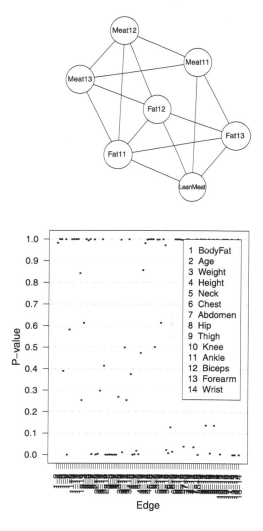

Fig. 4.13 A plot of the simultaneous *p*-values for the `gRbodyfat` data

The plot, which is shown in Fig. 4.13, is a lot less clear. If we let $\alpha = 0.1$, BodyFat is only connected to Abdomen. For $\alpha = 0.3$, also Wrist gets included in the predictor, as shown in Fig. 4.14.

```
> gsin.body <- as(getgraph(psin.body, 0.3), "graphNEL")
> plot(gsin.body, "neato")
```

This model is considerably sparser than the previously selected, with only 33 edges present:

```
> graph::degree(gsin.body)
```

BodyFat	Age	Weight	Height	Neck	Chest	Abdomen	Hip
2	5	8	5	2	5	7	6
Thigh	Knee	Ankle	Biceps	Forearm	Wrist		
7	3	4	2	5	5		

Fig. 4.14 Model for the
gRbodyfat data found by
thresholding the simultaneous
p-values

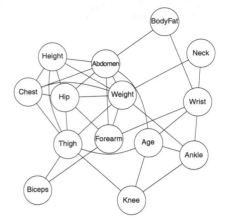

and it is probably too simple:

```
> sin.body <- cmod(gsin.body,data=gRbodyfat)
> sin.body$fitinfo$dev
```

```
[1] 244.6
```

```
> sin.body$fitinfo$dimension[4]
```

```
df
58
```

However, we should remember that the ambition of the SIN procedure is to identify
a graph that is a subgraph of the true graph, rather than the true graph itself, so it
attempts to only include edges that are definitely present.

4.4.5 Summary of Models

To get an overview of all the graphs selected by the various methods we could, for
example, look at the edges which are included in all of them.

```
> commonedges.carc<-intersection(as(aic.carc,"graphNEL"),
+ as(bic.carc,"graphNEL"))
> othermodels<-list(test.carc,forw.carc,
+                   thresh.carc,gsin.carc)
> othermodels<-lapply(othermodels, as, "graphNEL")
> for(ii in 1:length(othermodels))
+ {
+ commonedges.carc<-intersection(commonedges.carc,othermodels[[ii]])
+ }
> plot(commonedges.carc,"fdp")
```

Figure 4.15 shows the resulting graph. We note that all models selected have Fat11,
Fat12, Fat13, and LeanMeat in the same clique, i.e. indicate that LeanMeat is
directly associated to all fat measurements.

Fig. 4.15 The intersection of
the previously selected
models for the `carcass` data

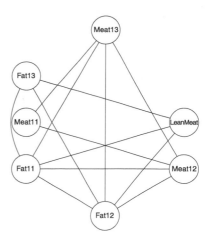

Fig. 4.16 The intersection of
the previously selected
models for the `gRbodyfat`
data

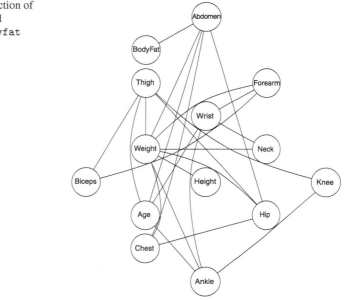

Figure 4.16 displays the corresponding graph for the `gRbodyfat` data. It appears
that the only stable neighbour of `BodyFat` is `Abdomen`:

```
> commonedges.body<-intersection(as(bic.body,"graphNEL"),
+ as(glasso.body,"graphNEL"))
> commonedges.body<-intersection(commonedges.body,gsin.body)
> plot(commonedges.body,"fdp")
```

4.5 Directed Gaussian Graphical Models

In the next two sections we consider models for Gaussian data that can be represented as directed acyclic graphs (DAGs). Recall that a probability distribution factorizes w.r.t. a DAG $\mathcal{G}$ if it can be expressed as

$$f(x) = \prod_{v \in V} f(x_v | x_{\mathrm{pa}(v)}) \qquad (4.18)$$

that is, as a product of conditional densities of individual variables given their parents in $\mathcal{G}$. To construct models of this type we need a list of univariate conditional models, one for each variable in V. In general, we may use any such models—linear or non-linear, additive or non-additive—but here we only consider linear regression models with Gaussian errors, since these are closely related to the undirected graphical Gaussian models described above. We call these *directed Gaussian graphical models*, or *DGGMs* for short.

The function `fitDag` in the **ggm** package fits DGGMs. For example,

```
> library(ggm)
> gdag1 <- DAG(LeanMeat ~ Meat13:Fat11:Fat12, Meat13 ~ Meat11:Meat12,
+            Fat12~Fat11, Fat13 ~ Meat11:Meat12, Meat12 ~ Meat11)
> plot(as(gdag1, "graphNEL"))
> fdag1 <- fitDag(gdag1, S.carc, nrow(carcass))
> fdag1$dev

[1] 552.3

> fdag1$df

[1] 12
```

The DAG is shown in Fig. 4.17.

Fig. 4.17 A DGGM for the carcass data

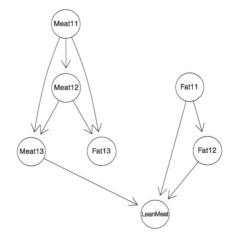

Fig. 4.18 The DGGMs (i) to
(iii) are distributionally
equivalent to the UGGM (iv)

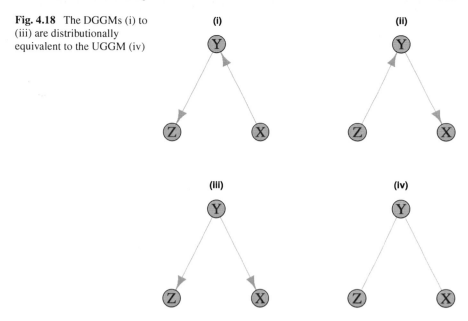

4.5.1 Markov Equivalence

Consider the graph in Fig. 4.18(iv). Since $(\{X\}, \{Y\}, \{Z\})$ is a decomposition of $\Gamma = \{X, Y, Z\}$, the joint density can be factorized into

$$f(x, y, z) = f(x) f(y|x) f(z|y)$$

where $f(y|x)$ and $f(z|y)$ are linear regressions. Thus the density factorizes according to the DAG shown in Fig. 4.18(i). Similarly, Fig. 4.18(i)–(iii) represent DGGMs that essentially are reparametrizations of the undirected graphical Gaussian model in Fig. 4.18(iv). In all four cases one conditional independence relation holds: that $X \perp\!\!\!\perp Z|Y$.

DAGs which induce the same sets of conditional independence relations are called *Markov equivalent*. Frydenberg (1990a) and Verma and Pearl (1990) showed that two DAGs are Markov equivalent if and only if they have the same skeleton and the same immoralities. The *skeleton* of a DAG is the undirected graph formed by replacing all arrows with (undirected) lines. An *unshielded collider* or *immorality* occurs when two directed edges from non-adjacent nodes meet head-on. For example, the DAGs shown in Fig. 4.18(i)–(iii) have the same skeleton and no unshielded colliders, and so are Markov equivalent. Although these models are *distributionally* equivalent, they are of course quite distinct when interpreted *causally*; we refrain from discussing this aspect further and refer to Spirtes et al. (1993) or Pearl (2000).

Since DGGMs that are Markov equivalent cannot be distinguished on the basis of sample distributions, model selection algorithms based on data samples can only select equivalence classes of DGGMs, not individual DGGMs. So it is important to

Fig. 4.19 A DAG with four
nodes and one immorality

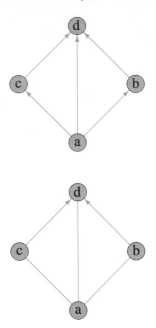

Fig. 4.20 The pDAG of the
DAG shown in Fig. 4.19

be able to represent these equivalence classes. There are several ways to do this, of
which we now describe two. Consider the DAG shown in Fig. 4.19:

This graph has one immorality, $b \rightarrow d \leftarrow c$. We can represent the equivalence
class using the graph constructed from the skeleton by orienting all edges that take
part in an immorality. This is called the *partially directed acyclic graph* or *pDAG*
for short (Chickering 2002). Figure 4.20 shows the pDAG for the current example.

Alternatively, we can construct a graph from the DAG by orienting all edges
whose direction is fixed in the equivalence class, and letting edges be undirected
if there are two members of the equivalence class which have arrows in opposite
directions. This construction has been given different names: the most commonly
used are *CPDAG* which is short for *completed partial directed acyclic graph* (Chickering 2002), *essential graph* (Andersson et al. 1996) or *pattern* (Verma and Pearl
1990). In the current example, if we try to reverse the edge $a \rightarrow d$, all orientations
of the remaining two edges introduce cycles or immoralities. So all DAGs in the
equivalence class have the arrow $a \rightarrow d$, and the essential graph is that shown in
Fig. 4.21:

The function `essentialGraph()` from the **ggm** package returns the essential
graph of a DAG. For example, Fig. 4.22 displays the essential graph of the DAG
shown in Fig. 4.17.

```
> eG1 <- as(essentialGraph(gdag1),"igraph")
> V(eG1)$size <- 40
> E(eG1)$arrow.mode <- 2
> E(eG1)[is.mutual(eG1)]$arrow.mode <- 0
> plot(eG1, layout=layout.kamada.kawai)
```

Fig. 4.21 The essential
graph of the DAG shown in
Fig. 4.19

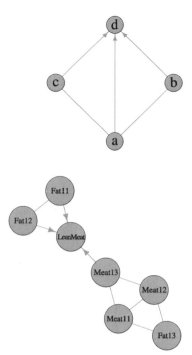

Fig. 4.22 The essential
graph of `gdag1`

The individual DAGs in the equivalence class can be obtained by assigning any orientation to the undirected edges, provided this does not introduce any cycles or immoralities into the graph. For example, orienting Meat13 → Meat11 and Fat13 → Meat11 in the above graph would introduce an immorality, and so is invalid.

To relate essential graphs to undirected graphs, consider a DAG generated by ordering the vertices in a triangulated graph using a perfect ordering. Then the DAG has no immoralities, and all DAGs generated in this way are Markov equivalent. Indeed, as any vertex can appear as the first in a perfect ordering, they constitute an equivalence class whose essential graph is the original triangulated graph. So a decomposable model can be regarded as being the essential graph of the class of DAGs generated from it using perfect orderings.

4.6 Model Selection for DGGMs

Sometimes variable orderings may be known in advance: for example, temporal orderings. In this case, model selection reduces to a series of univariate model selection tasks, regressing each variable on the prior variables. More often, however, variable ordering is not known in advance, and must be inferred from the data—but as we have seen, this can only be done up to Markov equivalence. In this section we illustrate some methods to do this.

4.6.1 The PC Algorithm

The *PC algorithm* (Spirtes and Glymour 1991; Spirtes et al. 1993) for model selection in DGGMs is implemented in the **pcalg** package. The algorithm has two
stages.

In the first stage the skeleton of the DAG is determined by exploiting the fact that
adjacency in the skeleton is given as

$$u \not\sim v \iff \exists S \subseteq V : u \perp\!\!\!\perp v | S.$$

Beginning from a complete graph, a series of conditional independence relations are
tested of successively increasing order and edges are removed as conditional independence relations are identified. That is, first marginal independences are tested,
then further relations of the form $u \perp\!\!\!\perp v | S$ for $|S| = 1, 2, \ldots$ and so on. To avoid
performing a huge number of independence tests, the PC algorithm exploits that at
any time, when an edge between u and v is tested, it is sufficient to consider sets S
which are subsets of $bd(u)$ or $bd(v)$. As edges are removed, the skeleton becomes
sparse, and the cardinality of S increases, such sets are very few.

This process results in a list of identified conditional independences, that is to
say, triplets (u, v, S) for which $u \perp\!\!\!\perp v | S$. The S sets are called sepsets, since they
correspond to sets which d-separate variables u and v in the unknown true DAG,
when such a DAG exists.

Comparing the skeleton and the dependence graph we note that the latter is defined by

$$u \not\sim v \iff u \perp\!\!\!\perp v | V \setminus \{u, v\}$$

so the skeleton for a DAG will typically have fewer edges than its independence
graph.

The following code illustrates the first stage of the algorithm as applied to the
carcass data. Note that we use pcalg::skeleton to ensure the relevant version
of the command skeleton is used.

```
> library(pcalg)
> C.carc<-cov2cor(S.carc)
> suffStat<-list(C=C.carc,n=nrow(carcass))
> indepTest<-gaussCItest
> skeleton.carc <- pcalg::skeleton(suffStat,gaussCItest,p=ncol(carcass),
+                                  alpha=0.05)
> nodes(skeleton.carc@graph)<-names(carcass)
```

The selected skeleton is shown in Fig. 4.23.

Note this has fewer edges than the dependence graphs found by model selection
among undirected models, suggesting that a DGGM might possibly be more appropriate. If indeed the data follow a DGGM, the undirected models in Sect. 4.2.1
may be interpreted as estimates of the moral graph of the DAG. The skeleton indicates that LeanMeat is directly associated to two of the fat measurements Fat11
and Fat12 and one of the meat measurements (Meat13).

Fig. 4.23 Skeleton selected
by PC algorithm for carcass
data

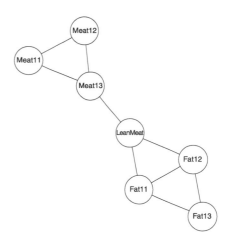

The second stage of the algorithm assigns directions to the edges in the skeleton, using the list of conditional independence relations found while establishing the skeleton. This information is contained in `skeleton.carc`. For example

```
> names(carcass)

[1] "Fat11"    "Meat11"   "Fat12"    "Meat12"   "Fat13"    "Meat13"
[7] "LeanMeat"

> str(skeleton.carc@sepset[[1]])

List of 7
 $ : NULL
 $ : int(0)
 $ : NULL
 $ : int(0)
 $ : NULL
 $ : int(0)
 $ : NULL
```

indicates that the first variable, `Fat11`, was marginally independent of variables `Meat11`, `Meat12`, and `Meat13`, leading to the corresponding edges being removed from the skeleton. Similarly

```
> str(skeleton.carc@sepset[[2]])

List of 7
 $ : NULL
 $ : NULL
 $ : int(0)
 $ : NULL
 $ : int 4
 $ : NULL
 $ : int 6
```

indicates that `Meat11` is marginally independent of `Fat12`, conditionally independent of `Fat13` given `Meat12`, and conditionally independent of `LeanMeat` given `Meat13`.

Fig. 4.24 The mixed graph
for the `carcass` data
returned by the PC-algorithm

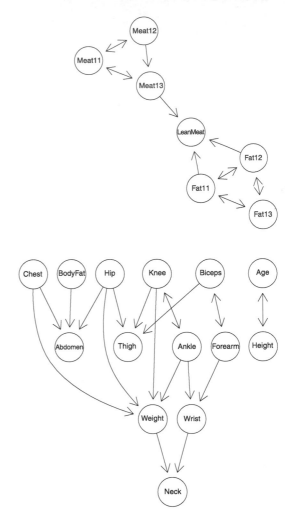

Fig. 4.25 The mixed graph
for the `gRbodyfat` data
returned by the PC-algorithm

Different algorithms are available for the second stage where directions are identified: in **pcalg** some of these are implemented in the functions udag2pdag(),
udag2pdagRelaxed() and udag2pdagSpecial(). See the help functions for an
explanation of these. The function udag2pdagRelaxed() here returns the mixed
graph shown in Fig. 4.24.

```
> pdag.carc <- udag2pdagRelaxed(skeleton.carc, verbose=0)
> nodes(pdag.carc@graph) <- names(carcass)
> plot(pdag.carc@graph,"neato")
```

Note that undirected edges are here represented as bi-directed edges; since
pdag.carc@graph is a graphNEL object, **Rgraphviz** is here used to plot it. Note
also that the algorithm did not succeed in returning a pDAG, since the edge from
Meat12 to Meat13 is not part of an immorality.

In practice it is easier to invoke both steps of the algorithm using the function pc. For the gRbodyfat data we could for example use:

```
> C.body<-cov2cor(S.body)
> suffStat.body<-list(C=C.body,n=nrow(gRbodyfat))
> cpdag.body <- pc(suffStat.body,gaussCItest,p=ncol(gRbodyfat),
                   alpha=0.01)
> nodes(cpdag.body@graph)<-names(gRbodyfat)
> plot(cpdag.body@graph)
```

which again identifies Abdomen as the central predictor for BodyFat: see Fig. 4.25.

The PC-algorithm gives a correct result (pDAG or essential graph) under the assumption that the distribution is *faithful* to a DAG and all independence relations are correctly decided. A probability distribution P is said to be faithful to a DAG $\mathcal{D}$ if all the conditional independences that hold under P can be inferred from $\mathcal{D}$ using the d-separation criterion.

However, if the true distribution is not faithful to a DAG, for example because it matches the conditional independence structure of a chordless four-cycle

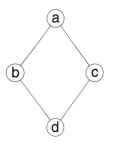

the algorithm will find in the first stage that $a \perp\!\!\!\perp d|\{b, c\}$ and $b \perp\!\!\!\perp c|\{a, d\}$, and correctly identify the skeleton as the given four-cycle; but no orientation of the edges is consistent with the conditional independence relations identified.

If, say, significance tests are used to decide conditional independences in the skeleton stage, it may happen by chance that the conditional independence relations may not be consistent with a DAG. In this case the second step is not guaranteed to return a pDAG, as we have just seen. Further inspection of the sepsets leads to the suspicion that a strong linear relation between the meat measurements is the culprit. A reasonable modification of the model is the DAG model shown in Fig. 4.26.

```
> gdag2 <- DAG(LeanMeat ~ Meat13:Fat11:Fat12, Meat13 ~ Meat11:Meat12,
+ Meat12 ~ Meat11, Fat11~Fat12:Fat13, Fat12~Fat13)
> plot(as(gdag2, "graphNEL"))
```

This, however, does not fit too well:

```
> fitDag(gdag2, S.carc, nrow(carcass))[c("dev","df")]
$dev
[1] 87.74

$df
[1] 12
```

Fig. 4.26 A DGGM for the
carcass data obtained by
modifying the output of the
PC-algorithm

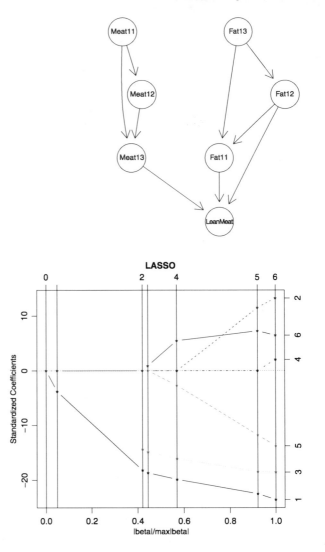

Fig. 4.27 The progress of the
lasso algorithm

This model implies that the meat measurements are marginally independent of the
fat measurements, with LeanMeat now depending on Fat11, Fat12, and Meat13.

For comparison we can look at a lasso regression of LeanMeat on the remaining
variables, using the lars() function in the **lars** package:

```
> library(lars)

Loaded lars 0.9-8

> lassoreg<-lars(as.matrix(carcass[c(1,2,3,4,5,6)]),
                 as.vector(carcass[7]))
> lassoreg

Call:
lars(x = as.matrix(carcass[c(1, 2, 3, 4, 5, 6)]),
```

```
      y = as.vector(carcass[7]))
R-squared: 0
Sequence of LASSO moves:
     Fat11 Fat12 Meat13 Fat13 Meat11 Meat12
Var      1     3      6     5      2      4
Step     1     2      3     4      5      6
```

which identifies the same variables as the three most important predictors, although Fat13 is next included. The progress of the algorithm is shown in Fig. 4.27.

```
> plot(lassoreg)
```

4.6.2 Alternative Methods for Identifying DGGMs

In the following sections we briefly illustrate some functions in the **bnlearn** package for selecting DGGMs. Note that these can also be used to select discrete DAG models, see Sect. 3.4.

4.6.2.1 Greedy Search

The hill-climbing algorithm is implemented in the hc function in the **bnlearn** package. This implements greedy search to optimize a score, for example the BIC. By this is meant that the current DAG is compared to all DAGs obtained by adding an edge, removing an edge, or reversing the direction of an edge. The model with the optimal score is chosen, and the process repeats until no score improvement can be made.

A potential problem with this approach is that the algorithm may get trapped in an equivalence class, since edge reversals within the class will not change the score. To continue the hill-climbing analogy: we may arrive at a ledge, and the way up is at the other end of the ledge, but we cannot move to the other end since we can only climb upwards. A method to deal with this is to use *random restarts*. When a local optimum is found, a pre-specified number of edges are perturbed (added/removed/reversed), and if the resulting model has improved score, then the process restarts at this new model.

We illustrate this process, starting off from the graph found by modifying the result of the PC algorithm.

```
> library(bnlearn)
> bn.init = empty.graph(nodes = names(carcass))
> amat(bn.init) = as(gdag2, "matrix")
```

Now we call the hc function, using bn.init as start model. If we omitted a start model, the null model would be used. The default score is the BIC. We specify 10 restarts, in which 4 edges are perturbed. In general hc may randomly return a cyclic graph which we have avoided by setting the seed of the random number generator.

Fig. 4.28 A DGGM for the
carcass data found using
the hill-climbing algorithm

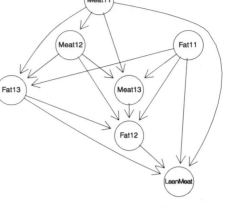

Fig. 4.29 The essential
graph of the DGGM found
previously

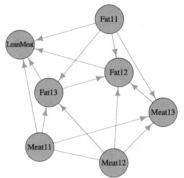

```
> set.seed(100)
> for (i in 1:6) carcass[,i] <- as.numeric(carcass[,i])
> bn.hc <- hc(carcass, bn.init, restart=10, perturb=4)
> fitDag(amat(bn.hc), S.carc, nrow(carcass))[c("dev","df")]

$dev
[1] 9.408

$df
[1] 6

> plot(as(amat(bn.hc),"graphNEL"), main="hill-climbing")
```

The model is shown in Fig. 4.28. The essential graph, shown in Fig. 4.29, is obtained
as follows:

```
> eg.hc <- as(essentialGraph(amat(bn.hc)),"igraph")
> E(eg.hc)$arrow.mode <- 2
> E(eg.hc)[is.mutual(eg.hc)]$arrow.mode <- 0
> plot(eg.hc, layout=layout.kamada.kawai, vertex.size=40)
```

An attractive feature of the model selected by the PC algorithm was that the
LeanMeat variable was a response to the fat and meat variables. This is in accor-

Fig. 4.30 A DGGM for the
`carcass` found using
hill-climbing with a blacklist

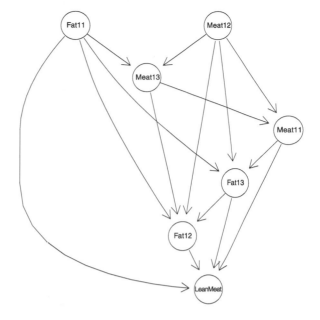

dance with expectation, since `LeanMeat` is essentially derived from the fat and meat
variables. If we wish to build on this prior belief, we can disallow all edges from
`LeanMeat` to the other variables by specifying a blacklist, that is, a list of forbidden
edges.

```
> bl <- data.frame(from=names(carcass)[rep(7,6)],
                   to=names(carcass)[1:6])
> bn.hcbl <- hc(carcass, bn.init, restart=20, perturb=4, blacklist=bl)
> fitDag(amat(bn.hcbl), S.carc, nrow(carcass))[c("dev","df")]

$dev
[1] 7.894

$df
[1] 6

> plot(as(amat(bn.hcbl),"graphNEL"))
```

The selected model is shown in Fig. 4.30. We should note that also the PC algorithm itself can take constraints in the form of edges which are known to be present,
absent, or having fixed directions.

4.6.2.2 A Hybrid Algorithm

We now illustrate the use of a hybrid constraint/score-based algorithm, the max–
min hill-climbing algorithm of Tsamardinos et al. (2003), implemented in the `mmhc`
function in the **bnlearn** package. A constraint-based algorithm is used to find the
skeleton, which is then oriented using a greedy hill-climbing algorithm. Per default,

Fig. 4.31 A DGGM for the
carcass data found using
max–min hill-climbing

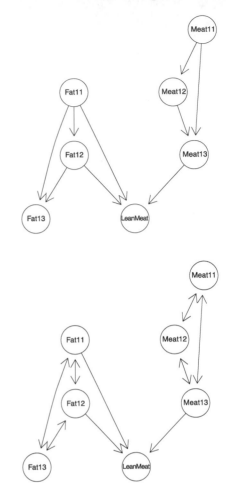

Fig. 4.32 The essential
graph of the DGGM found
previously

the first step uses significance tests with level 0.05, and the second step uses BIC
scores.

```
> bn.mmhc <- mmhc(carcass)
> fitDag(amat(bn.mmhc), S.carc, nrow(carcass))[c("dev","df")]

$dev
[1] 87.74

$df
[1] 12

> plot(as(amat(bn.mmhc),"graphNEL"))
```

The selected model is shown in Fig. 4.31. We note that this is the same model as we
originally found by modifying the result of the PC algorithm. The essential graph is
shown in Fig. 4.32.

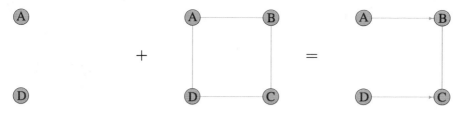

Fig. 4.33 Constructing a chain graph model

```
> plot(as(essentialGraph(amat(bn.mmhc)),"graphNEL"))
```

If we compare the models found by the different algorithms we find that the constraint-based and hybrid algorithms (the PC algorithm and the max–min hill-climbing algorithm) tend to find simple models with relatively poor overall fit, conforming with the fact that they are not optimizing a global score. In contrast, the score-based algorithms (the hill-climbing algorithm, with or without blacklisting) find more complex models that overall fit relatively well.

4.7 Gaussian Chain Graph Models

In this section we briefly consider models for Gaussian data that can be represented as chain graphs.

The factorization requirements were described in Sect. 1.3. For each chain component $C \in \mathcal{C}$ in the component DAG, the conditional densities $f(x_C | x_{\mathrm{pa}(C)})$ must factorize according to an undirected graph formed by moralizing the graph induced by $C \cup \mathrm{pa}(C)$.

A convenient way of constructing such models is to derive them from the corresponding undirected models by conditioning. For example, the chain graph model shown on the right of Fig. 4.33 is constructed by combining the marginal model for (A, D) (shown on the left) and the conditional model for $c = \{B, C\}$ given (A, D) induced by the undirected model shown in the middle. The conditional distributions are found in the usual way from the joint distribution under the model: see (4.11).

4.7.1 Selecting a Chain Graph Model

Several functions in the **lcd** package enable a general Gaussian chain graph model to be selected, using an algorithm due to Ma et al. (2008). This is a constraint-based algorithm that consists of three steps:

1. Firstly, an undirected graphical model for the data is chosen. Any conditional independences that hold under this model will also hold under the selected chain graph, so this step serves to restrict the search space in the third step.

Fig. 4.34 A chain graph model for the `carcass` data found using the `lcd` algorithm

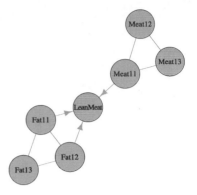

2. A junction tree for the undirected graph is derived.
3. The algorithm proceeds by performing a series of conditional independence tests, following a scheme based on the junction tree. Provided that the test results are consistent with a chain graph, the algorithm is guaranteed to return this.

We illustrate this process on the carcass data.

```
>    library(lcd)
>    ug <- naive.getug.norm(carcass, 0.05)
>    jtree <- ug.to.jtree(ug)
>    cg <- learn.mec.norm(jtree, cov(carcass), nrow(carcass), 0.01,
                          "CG")
>    icg <- as(cg, "igraph")
>    E(icg)$arrow.mode <- 2
>    E(icg)[is.mutual(icg)]$arrow.mode <- 0
>    V(icg)$size <- 40
>    plot(icg, layout=layout.kamada.kawai)
```

Here we use a simple thresholding algorithm in the **lcd** package to find the undirected model for the first step. Any of the algorithms described in Sect. 4.4 could in principle be used here. We note that the selected chain graph is very similar to that selected by the max–min hill-climbing algorithm above. The chain graph selected, shown in Fig. 4.34, is in fact Markov equivalent to the graph we found by modifying the output of the PC algorithm, shown in Fig. 4.26, and equal to the essential graph of that DAG.

4.8 Various

The package **ggm** has facilities for working with other classes of graphical models for Gaussian data, for example covariance graph models and ancestral graph models. In addition to those described in this chapter, a number of packages support model selection for Gaussian graphical models, including **qp**, **GeneNet** and **gRapHD** (see Chap. 7).

Chapter 5
Mixed Interaction Models

5.1 Introduction

This chapter introduces *mixed interaction models*, a class of models for discrete and continuous variables that combine log-linear models for discrete variables (described in Chap. 2) with graphical Gaussian models for continuous variables (described in Chap. 4). The exposition given here is restricted to *homogeneous mixed interaction models*. Homogeneity in this context means that the covariance matrix of the Gaussian variables does not depend on the values of discrete variables. More general types of mixed interaction models that do not assume homogeneity are described in Lauritzen (1996) and Edwards (2000). An important advantage of the homogeneous models is that they can be specified using model formulae that are similar to the model formulae for log-linear models and for graphical Gaussian models.

5.2 Example Datasets

To introduce the models we consider three datasets that are in **gRbase**. The first dataset, milkcomp1, comes from a study comparing the composition of sow milk in terms of fat, protein and lactose content under 8 different diets. The control diet consisted of soybean meal, barley and wheat. The other diets added 8% fat to this basis diet: animal fat, rapeseed oil, fish oil, coconut oil, palm oil or sunflower oil. Sow milk was analysed for the concentration of dry matter, protein, fat and lactose: here we consider the data recorded four days after farrowing (i.e., giving birth). For further details see Lauridsen and Danielsen (2004). The first rows of the dataset are:

```
> data(milkcomp1, package='gRbase')
> head(milkcomp1)

  treat  fat protein    dm lactose
1     d 6.16    6.65 18.55    5.06
2     c 4.06    5.44 18.32    5.23
3     f 9.25    5.67 20.68    5.15
```

S. Højsgaard et al., *Graphical Models with R*, Use R!,
DOI 10.1007/978-1-4614-2299-0_5, © Springer Science+Business Media, LLC 2012

```
4       b 5.82    5.62 17.57    5.74
5       a 4.98    5.37 16.38    5.55
6       b 9.06    5.08 20.21    5.29
```

The second dataset, `wine`, contains the results of a study of the chemical constituents of three varieties of grape, grown in the same region in Italy. There are 178 observations on 14 variables, of which one is discrete (grape variety) and the rest (chemical constituents) are continuous. For more information on this dataset see http://archive.ics.uci.edu/ml/datasets/Wine.

```
> data(wine, package='gRbase')
> head(wine)

  Cult  Alch Mlca  Ash Aloa Mgns Ttlp Flvn Nnfp Prnt Clri  Hue Oodw
1   v1 14.23 1.71 2.43 15.6  127 2.80 3.06 0.28 2.29 5.64 1.04 3.92
2   v1 13.20 1.78 2.14 11.2  100 2.65 2.76 0.26 1.28 4.38 1.05 3.40
3   v1 13.16 2.36 2.67 18.6  101 2.80 3.24 0.30 2.81 5.68 1.03 3.17
4   v1 14.37 1.95 2.50 16.8  113 3.85 3.49 0.24 2.18 7.80 0.86 3.45
5   v1 13.24 2.59 2.87 21.0  118 2.80 2.69 0.39 1.82 4.32 1.04 2.93
6   v1 14.20 1.76 2.45 15.2  112 3.27 3.39 0.34 1.97 6.75 1.05 2.85
  Prln
1 1065
2 1050
3 1185
4 1480
5  735
6 1450
```

The third dataset, `Nutrimouse`, stems from a study of the effect of nutrition on lipid levels and gene expression in mice. Forty mice were each assigned one of five different diets, with different fatty acid compositions. Two strains of mice were used, one with the PPARα gene knocked out and the other was wild-type (i.e. the PPARα gene was present). The PPARα gene is known to affect fatty acid metabolism. The concentrations of 21 lipids (fatty acids) in the liver were recorded. In addition the data include the expression levels of 120 genes in the liver: these 120 were selected from a much greater number as potentially relevant for nutrition. Thus the dataset contains $N = 40$ observations of 143 variables: two discrete design variables—genotype (with two levels) and diet (with five levels), 120 gene expression values and 21 lipid values. For more details see Martin et al. (2007).

The following code fragment lists a small subset of the data.

```
> data(Nutrimouse, package='gRbase')
> head(Nutrimouse[,c(1:5,123:126)])

  genotype diet X36b4 ACAT1 ACAT2 C14.0 C16.0 C18.0 C16.1n.9
1       wt  lin -0.42 -0.65 -0.84  0.34 26.45 10.22     0.35
2       wt  sun -0.44 -0.68 -0.91  0.38 24.04  9.93     0.55
3       wt  sun -0.48 -0.74 -1.10  0.36 23.70  8.96     0.55
4       wt fish -0.45 -0.69 -0.65  0.22 25.48  8.14     0.49
5       wt  ref -0.42 -0.71 -0.54  0.37 24.80  9.63     0.46
6       wt  coc -0.43 -0.69 -0.80  1.70 26.04  6.59     0.66
```

5.3 Mixed Data and CG-densities

Suppose that N observations of d discrete variables and q continuous variables are available. We denote the set of discrete variables by Δ, the set of continuous variables by Γ, and the combined variable set by $V = \Delta \cup \Gamma$.

An observation has the form $x = (i, y) = (i_1, \ldots, i_d, y_1, \ldots, y_q)$. This combines the notation of Chap. 2 and Chap. 4. As in Chap. 2 we write the set of possible cells $i = (i_1, \ldots, i_d)$ as $\mathcal{I}$.

We construct a homogeneous *conditional Gaussian density*, or *CG-density* for short, for $x = (i, y)$ in the following way. Firstly, the probability of the discrete variables falling in cell i is denoted $p(i)$. We assume that $p(i) > 0$ for all $i \in \mathcal{I}$. Secondly, the conditional distribution of the continuous variables given that the discrete variables fall in cell i is multivariate Gaussian $\mathcal{N}\{\mu(i), \Sigma\}$. Observe that the mean may depend on i but the variance does not. The density takes the form

$$f(i, y) = p(i)(2\pi)^{-q/2} \det(\Sigma)^{-1/2} \exp\left[-\frac{1}{2}\{y - \mu(i)\}^{\top} \Sigma^{-1}\{y - \mu(i)\} \right] \quad (5.1)$$

The parameters $\{p(i), \mu(i), i \in \mathcal{I}; \Sigma\}$, that is, the cell probability and mean vector for each cell i and the common covariance matrix, are called the *moment parameters*.

It is convenient to represent (5.1) in *exponential family* form as

$$f(i, y) = \exp\left\{ g(i) + \sum_u h^u(i) y_u - \frac{1}{2} \sum_{uv} y_u y_v k_{uv} \right\}$$

$$= \exp\left\{ g(i) + h(i)^{\top} y - \frac{1}{2} y^{\top} K y \right\} \quad (5.2)$$

The parameters $\{g(i), h(i), i \in \mathcal{I}; K\}$ are called the *canonical parameters*. Note that the canonical parameters have the same dimensions as the moment parameters: for each i, $g(i)$ is a scalar (the discrete canonical parameter) and $h(i)$ is a q-vector (the linear canonical parameter); also, the *concentration matrix* K is a symmetric positive definite $q \times q$ matrix.

Occasionally it is convenient to use the *mixed parameters* which are given as $\{p(i), h(i), i \in \mathcal{I}; K\}$. We allow ourselves to write the parameters briefly as $\{p, \mu, \Sigma\}$, $\{g, h, K\}$ and $\{p, h, K\}$.

We can transform back and forth between the different parameterizations using the relations

$$K = \Sigma^{-1}$$

$$h(i) = \Sigma^{-1}\mu(i)$$

$$g(i) = \log p(i) - \frac{1}{2}\log\det(\Sigma) - \frac{1}{2}\mu(i)^{\top}\Sigma^{-1}\mu(i) - \frac{q}{2}\log(2\pi),$$

and

$$\Sigma = K^{-1}$$

$$\mu(i) = K^{-1}h(i)$$

$$p(i) = (2\pi)^{\frac{q}{2}} \det(K)^{-\frac{1}{2}} \exp\left\{g(i) + \frac{1}{2}h(i)^{\top} K^{-1}h(i)\right\}. \qquad (5.3a)$$

5.4 Homogeneous Mixed Interaction Models

The *homogeneous mixed interaction models*, which we for brevity here refer to as *MI-models*, are defined by constraining the canonical parameters of CG-densities so as follow factorial expansions.

For example, let $\Delta = \{A, B\}$ and $\Gamma = \{X, Z\}$ and let the levels of the factors A and B be denoted j and k. So in this case $i = (j, k)$ and $y = (x, z)$. The joint density can be written

$$f(i, y) = \exp\left\{g(i) + h^x(i)x + h^z(i)z - \frac{1}{2}(k_{xx}x^2 + 2k_{xz}xz + k_{zz}z^2)\right\} \qquad (5.4)$$

and we can write the unrestricted (or saturated) model as

$$g(i) = u + u_j^a + u_k^b + u_{jk}^{ab} \qquad (5.5)$$

$$h^x(i) = v + v_j^a + v_k^b + v_{jk}^{ab} \qquad (5.6)$$

$$h^z(i) = w + w_j^a + w_k^b + w_{jk}^{ab} \qquad (5.7)$$

$$K = \begin{pmatrix} k_{xx} & k_{xz} \\ k_{xz} & k_{zz} \end{pmatrix} \qquad (5.8)$$

where the u's, v's and w's are interaction terms. In this model $g(i)$, $h^x(i)$ and $h^y(i)$ are unrestricted functions of the cells $i = (j, k)$. To estimate the interaction terms uniquely would require some further constraints but we do not bother about this here. This is because we use the *factorial expansions* to constrain the way canonical parameters vary over $\mathcal{I}$, but are not usually interested in their values *per se*.

Models are defined by setting certain interaction terms to zero. The usual hierarchical rule, that if a term is set to zero then all higher-order terms must also be zero, is respected. So by this rule, if we set v_j^a to zero for all j, we must also set v_{jk}^{ab} to zero for all j and k.

Conditional independence constraints can be imposed by setting interaction terms to zero. For example, to make $A \perp\!\!\!\perp X \mid (B, Z)$ we must set all terms involving A and X in (5.4) to zero, that is, $v_j^a = v_{jk}^{ab} = 0$, $\forall j, k$. To make $A \perp\!\!\!\perp B \mid (X, Z)$ we must set all terms involving A and B to zero, i.e., $u_{jk}^{ab} = v_{jk}^{ab} = w_{jk}^{ab} = 0$, $\forall j, k$. Finally, to obtain $X \perp\!\!\!\perp Z \mid (A, B)$ we set $k_{xz} = 0$.

For example, consider the `milkcomp1` data:

```
> head(milkcomp1,3)
  treat  fat protein    dm lactose
1     d 6.16    6.65 18.55    5.06
2     c 4.06    5.44 18.32    5.23
3     f 9.25    5.67 20.68    5.15
```

The `CGstats()` function calculates the number of observations and the means of the continuous variables for each cell i, as well as (by default) a common covariance matrix:

```
> library(gRim)
> SS <- CGstats(milkcomp1, varnames=c("treat","fat","protein",
                                        "lactose"))
> SS

$n.obs
treat
a b c d e f g
8 8 8 8 8 7 8

$center
            a     b     c     d     e     f     g
fat     6.641 8.010 7.053 7.401 8.134 7.519 6.974
protein 5.487 5.287 5.475 5.817 5.263 5.296 5.580
lactose 5.491 5.489 5.468 5.314 5.406 5.383 5.415

$cov
            fat protein lactose
fat     2.31288 0.19928 -0.07028
protein 0.19928 0.12289 -0.03035
lactose -0.07028 -0.03035  0.04530
```

Note that the mean of fat (and to a lesser extent of protein) varies over the treatments whereas the lactose means seem to be more or less constant. The coefficients of variation are:

```
> apply(SS$center,1,sd) / apply(SS$center,1,mean)
    fat protein lactose
0.07416 0.03656 0.01187
```

The corresponding canonical parameters are

```
> can.parms<-CGstats2mmodParms(SS,type="ghk")
> print(can.parms, simplify=FALSE)

MIparms: form=ghk
$g
treat
     a      b      c      d      e      f      g
-745.5 -729.4 -740.5 -743.6 -712.7 -710.5 -740.2

$h
           a       b        c        d       e       f        g
[1,]   0.787   1.628   0.9976   0.8736   1.686   1.344   0.8642
[2,]  88.221  85.006  87.6318  90.1511  84.137  84.817  88.5107
[3,] 181.555 180.651 180.9626 179.0642 178.338 177.745 180.1856
```

```
$K
         [,1]      [,2]      [,3]
[1,]    0.5056  -0.7503   0.2817
[2,]   -0.7503  10.8649   6.1158
[3,]    0.2817   6.1158  26.6104
```

Let j refer to a level of the treatment factor. Then $h(j)$ takes the form

$$h(j) = (h^{\text{fat}}(j), h^{\text{protein}}(j), h^{\text{lactose}}(j)).$$

The coefficients of variation for h are

```
> apply(can.parms$h,1,sd) / apply(can.parms$h,1,mean)
```

`[1] 0.324840 0.026150 0.007934`

which suggests that $h^{\text{lactose}}(j)$ is constant as a function of j; that is

$$h(j) = (h^{\text{fat}}(j), h^{\text{protein}}(j), h^{\text{lactose}}).$$

If we insert this in (5.2) and use the factorization criterion 1.1 we find that

$$\text{lactose} \perp\!\!\!\perp \text{treat} \,|\, (\text{fat}, \text{protein}).$$

The partial correlation matrix is more informative than the concentration matrix:

```
> conc2pcor(can.parms$K)
```

```
          [,1]      [,2]       [,3]
[1,]   1.00000   0.3201   -0.07679
[2,]   0.32014   1.0000   -0.35968
[3,]  -0.07679  -0.3597    1.00000
```

This suggests that the partial correlation between `fat` and `lactose` is zero. If we set $k_{\text{fat,lactose}} = 0$ in (5.2) and use the factorization criterion we find that

$$\text{lactose} \perp\!\!\!\perp \text{fat} \,|\, (\text{treat}, \text{protein}).$$

5.5 Model Formulae

In this section we describe how to specify MI-models using *model formulae* and show how they may be represented as *dependence graphs*. Here and later we refer to the models and graphs shown in Table 5.1.

As we have seen above in Sect. 5.4, we define an MI-model by constraining $g(i)$ and the $h^u(i)$ for $u \in \Gamma$ to satisfy factorial expansions, and by constraining some off-diagonal elements of K to zero. So in principle we can define an MI-model by giving a list of generating classes—one for $g(i)$ and one for $h^u(i)$ for each $u \in \Gamma$—together with list of off-diagonal elements of K that are allowed to be non-zero. Together these specifications define an MI-model, although some restrictions in the different components are necessary, as we describe below.

Table 5.1 Some homogeneous mixed interaction models

Model	Formula	Graph	Graphical	Decomposable
(a)	A*B*X*Z		true	true
(b)	A*B*Z+B*X*Z		true	true
(c)	A*B*Z+A*X		true	true
(d)	A*Z+B*Z+A*X		true	false
(e)	A*X+A*Z+B*X+B*Z		true	false
(f)	A*B+A*Z+B*X*Z		false	false
(g)	A*X+B*X		true	false

To give all these generating classes would be very cumbersome, however. It is much more convenient to specify a model using a single generating class $\mathcal{C} = \{G_1, \ldots, G_m\}$, with $G_j \subseteq V$ for each $j = 1 \ldots m$. We now explain how this is done.

We use the following convention. We write a generator G as a pair (a, b) where $a = G \cap \Delta$ are discrete variables and $b = G \cap \Gamma$ are continuous variables. For $a \subset \Delta$, by $g_a(i_a)$ we mean a function which depends on an index i only through i_a. Let q be the number of variables in Γ. Suppose that y is a q-vector. For $b \subset \Gamma$ we write the corresponding subvector of y as y^b. Furthermore, we take $[y^b]$ to mean the q-vector obtained by padding y^b with zeros in the right places to obtain full dimension.

Using this convention we can define the restrictions which a generating class $\mathcal{C}$ imposes on a general (homogeneous) CG-density.

1. The discrete canonical parameter $g(i)$ is constrained to follow the factorial expansion

$$g(i) = \sum_{(a,b) \in \mathcal{C}} g_a(i_a)$$

That is to say, the generators for $g(i)$ are the maximal elements of $\{a \mid (a, b) \in \mathcal{C}\}$, which we write compactly as $\max(\{a \mid (a, b) \in \mathcal{C}\})$. These are called the *discrete generators* of the model.

2. The linear canonical parameter h is constrained to follow the factorial expansion

$$h(i) = \sum_{(a,b) \in \mathcal{C}} [h_a^b(i_a)].$$

It follows that $h(i)^\top y = \sum_{(a,b) \in \mathcal{C}} h_a^b(i_a)^\top y_b$. For each $u \in \Gamma$, the generators for $h^u(i)$ are $\mathcal{C}^u = \max(\{a \mid (a,b) \in \mathcal{C} \wedge u \in b\})$; that is, the discrete components of those generators containing u. These are termed the *linear generators* of the model.

3. Finally, the quadratic canonical parameter K is constrained as follows: elements k_{uv} of K are set to zero unless $\{u,v\} \subset b$ for some generator $(a,b) \in \mathcal{C}$. The sets $\{b \mid (a,b) \in \mathcal{C}\}$ induce a graph whose edges of correspond to those k_{uv} which are not set to zero. The cliques of the graph are called the *quadratic generators* of the model.

For example, the last model in Table 5.1 has the generating class

$$\{(A, B), (A, Z), (B, X, Z)\}.$$

The derived formulae for $g(i)$, $h^x(i)$ and $h^z(i)$ are $\{(A, B)\}$, $\{(B)\}$, and $\{(A), (B)\}$, respectively. Hence $g(i)$ is unrestricted, $h^x(i)$ satisfies $h^x(i) = v + v_k^b$ for all $i = (j, k)$ and $h^z(i)$ satisfies $h^z(i) = w + w_j^a + w_k^b$ for all $i = (j, k)$. Since $(X, Z) \subset (B, X, Z)$, k_{xz} is not set to zero.

It can be shown that to ensure location and scale invariance, the formula for $g(i)$ must be "larger" than the formulae for each $h^u(i)$ in the sense that each generator for $h^u(i)$ must be contained in a generator for $g(i)$. This constraint is automatically fulfilled by the above construction.

The model formula notation for MI-models used here has the disadvantage that distinct formulae can specify the same model. For example, if $\Delta = \{I\}$ and $\Gamma = \{X, W, Z\}$ then the formulae I*X*W+X*W*Z and I*X*W+X*Z+W*Z give identical models. This is not usually problematic, but it can impact the efficiency of the iterative estimation procedure, as we describe later. We can define a particular representation, termed the *maximal form* of the model. This has generators defined as the maximal sets $\mathcal{A} \subseteq \Delta \cup \Gamma$ such that:

1. $\mathcal{A} \cap \Delta$ is contained in a generator of $g(i)$,
2. for each $u \in \mathcal{A} \cap \Gamma$, $\mathcal{A} \cap \Delta$ is contained in a generator of $h^u(i)$, and
3. for each $x, y \in \mathcal{A} \cap \Gamma$, with $u \neq v$, k_{uv} is not set to be zero.

For example, I*X*W+X*W*Z is of maximal form but I*X*W+X*Z+W*Z is not.

The mmod() function in the **gRim** package allows MI-models to be defined using model formulae. For example, to define the model for the milk composition dataset with the conditional independences arrived at in Sect. 5.4, we specify the generating class with generators {treat,fat,protein} and {protein,lactose}, as follows:

```
> milkmod <- mmod(~treat*fat*protein + protein*lactose, data=milkcomp1)
```

Fig. 5.1 Mixed interaction
model for milk composition
data. Discrete variables are
shown as *grey* nodes while
continuous variables are *white*

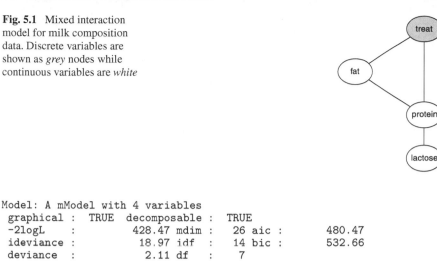

```
Model: A mModel with 4 variables
 graphical :  TRUE  decomposable :  TRUE
 -2logL    :       428.47 mdim :   26 aic :       480.47
 ideviance :        18.97 idf  :   14 bic :       532.66
 deviance  :         2.11 df   :    7
```

The discrete, linear and quadratic generators of the model are

```
> str(milkmod$modelinfo$dlq)

List of 3
 $ discrete :List of 1
  ..$ : chr "treat"
 $ linear    :List of 2
  ..$ : chr [1:2] "fat" "treat"
  ..$ : chr [1:2] "protein" "treat"
 $ quadratic:List of 2
  ..$ : chr [1:2] "fat" "protein"
  ..$ : chr [1:2] "protein" "lactose"
```

To construct the dependence graph of an MI-model defined using such a formula, we
connect with an edge all variable pairs appearing in the same generator. By convention, discrete variables are drawn with filled circles and continuous variables with
hollow circles. The *global Markov property* (Sect. 1.3) can be used for reading conditional independencies from the dependence graph in the usual way. For example,
the dependence graph for the model `milkmod` just discussed is shown in Fig. 5.1. It
can be obtained using the `plot` function:

```
> plot(milkmod)
```

5.6   Graphical and Decomposable MI-models

Suppose we are given an undirected graph with vertex set $\Delta \cup \Gamma$ and consider the
MI-model for $\Delta \cup \Gamma$ whose generators are the cliques of the graph. An MI-model
that can be formed in this way is termed a *graphical MI-model*. Table 5.1 shows
some graphical MI-models.

As with log-linear models, it is possible to set higher-order interactions to zero,
without introducing new conditional independence relations. Such models are called

non-graphical. For example, consider model (b) in Table 5.1. Since the generators of the formula correspond to the cliques of the graph, the model is graphical. The model implies that the term $h^y(i)$ is unrestricted, say as

$$h^y(i) = w + w_j^a + w_k^b + w_{jk}^{ab}.$$

If we constrain $w_{jk}^{ab} = 0$, $\forall j, k$, then $h^y(i)$ has the additive form $h^y(i) = w + w_j^a + w_k^b$, $\forall j, k$. This does not correspond to a conditional independence restriction, but results in model (f) in Table 5.1. So model (f) is non-graphical. Since no further conditional independence restrictions have been added model (f) has the same dependence graph as model (b).

We now turn to a subclass of the graphical MI-models, the *decomposable MI-models.* These build on a more basic concept, that of a *decomposition*, which we describe first.

The notion of a decomposition of a graph $\mathcal{G}$ with mixed variables relates to the question of how and when the analysis of a graphical MI-model may be broken down into analyses of smaller models. This notion is slightly more elaborate than in the purely discrete and purely continuous cases. Let A, B and S be disjoint non-empty subsets of V such that $A \cup B \cup S = V$. We define (A, B, S) to be a *decomposition* of $\mathcal{G}$ if the following conditions hold:

1. A and B are separated by S in $\mathcal{G}$,
2. S is complete in $\mathcal{G}$, and
3. $S \subset \Delta$ or $B \subset \Gamma$.

It can be shown that when (A, B, S) is a decomposition of $\mathcal{G}$, the maximum likelihood estimator $\hat{f}$ of the density of the graphical MI-model with dependence graph $\mathcal{G}$ is given by

$$\hat{f} = \frac{\hat{f}_{[A \cup S]} \hat{f}_{[B \cup S]}}{\hat{f}_{[S]}}$$

where $\hat{f}_{[A \cup S]}$, $\hat{f}_{[B \cup S]}$, $\hat{f}_{[S]}$ are the estimates of densities based on the models corresponding to the relevant induced subgraphs and based on marginal data only. Indeed they are weak marginals of $\hat{f}$, see Sect. 5.7.5.1 below.

A graph with mixed variables $\mathcal{G}$ is called decomposable if it is complete or it can be successively decomposed into complete graphs.

Various characterizations of graphs with this property are useful. One is based on the forbidden path property: a *forbidden path* is a path between two non-adjacent discrete vertices that passes through only continuous vertices. It can be shown that a graph is decomposable if and only if it is triangulated and has no no forbidden paths. A simple example of a graph with mixed variables that is not decomposable is:

Another characterization is that the cliques of a decomposable graph with mixed variables can be ordered as $(C_1, \ldots, C_k)$ with a modified version of the *running in-*

Fig. 5.2 Decomposable graphs with mixed variables. If a and d are discrete and b and c are continuous then the first graph is not decomposable whereas the second graph is

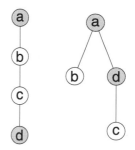

tersection property. For $j > 1$ define $H_j = \bigcup_{t=1}^{j-1} C_t$ and $S_j = C_j \cap H_j$. The modified condition is that

1. for each $j > 1$, $S_j \subset C_i$ for some $i < j$, and
2. for each $j > 1$ it holds that $C_j \setminus S_j \subseteq \Gamma$ or $S_j \subseteq \Delta$.

The additional condition (2) states that continuous variables cannot be prior to discrete ones. A graph with mixed variables is decomposable if and only there exists an ordering of its cliques fulfilling conditions (1) and (2).

A decomposable MI-model is a graphical MI-model whose dependence graph is decomposable. For such a model, the maximum likelihood estimates take the closed form

$$\hat{f}(x) = \prod_{j=1}^{k} \frac{\hat{f}_{[C_j]}(x_{C_j})}{\hat{f}_{[S_j]}(x_{S_j})} \qquad (5.9)$$

where we have let $S_1 = \emptyset$ and $\hat{f}_\emptyset = 1$.

To check whether a graph with mixed variables is decomposable, the so-called *star graph* construction can be used. That is, let $\mathcal{G}^\star$ be a new graph obtained by adding an extra vertex, $\star$, to $\mathcal{G}$ and adding edges between $\star$ and all discrete variables. Then $\mathcal{G}^\star$ is triangulated (which can be checked with maximum cardinality search) if and only if $\mathcal{G}$ is decomposable.

It can also be shown that a graph $\mathcal{G}$ with mixed variables is decomposable if and only if the vertices of $\mathcal{G}$ can be given a *perfect ordering*. For such graphs this is defined as an ordering $\{v_1, v_2, \dots, v_T\}$ such that (i) $S_k = \text{ne}(v_k) \cap \{v_1, v_2, \dots, v_{k-1}\}$ is complete in $\mathcal{G}$ and (ii) $S_k \subset \Delta$ if $v_k \in \Delta$. The mcsmarked() function is based on constructing $\mathcal{G}^\star$ as described above and returns a perfect ordering if the graph is decomposable.

As an example consider the following two graphs shown in Fig. 5.2. If a and d are discrete and b and c are continuous then the graph on the left is not decomposable whereas the graph on the right is. Note that since a graph object contains no information about whether the nodes are discrete or continuous, mcsmarked() has to be supplied this information explicitly.

```
> uG1 <- ug(~a:b+b:c+c:d)
> uG2 <- ug(~a:b+a:d+c:d)
> mcsmarked(uG1, discrete=c("a","d"))
```

```
character(0)
> mcsmarked(uG2, discrete=c("a","d"))
[1] "a" "d" "b" "c"
```

5.7 Maximum Likelihood Estimation

In this section we derive expressions for the likelihood and describe algorithm(s) the maximize this.

5.7.1 Likelihood and Deviance

In this section we derive some expressions for the *likelihood* and the *deviance*. The log density can be written as

$$\log f(i, y) = \log p(i) - q \log(2\pi)/2 - \log \det(\Sigma)/2$$
$$- \{y - \mu(i)\}^{\top} \Sigma^{-1} \{y - \mu(i)\}/2,$$

so the log-likelihood of a sample (i^{ν}, y^{ν}), $\nu = 1, \ldots, N$ is

$$\ell = \sum_{i} n(i) \log p(i) - Nq \log(2\pi)/2 - N \log \det(\Sigma)/2$$
$$- \sum_{\nu=1}^{N} \{y^{\nu} - \mu(i^{\nu})\}^{\top} \Sigma^{-1} \{y^{\nu} - \mu(i^{\nu})\}/2.$$

We can simplify the last term using that

$$\sum_{i} \sum_{\nu : i^{\nu} = i} \{y^{\nu} - \mu(i)\}^{\top} \Sigma^{-1} \{y^{\nu} - \mu(i)\}$$
$$= N \operatorname{tr}(S \Sigma^{-1}) + \sum_{i} n(i) \{\bar{y}(i) - \mu(i)\}^{\top} \Sigma^{-1} \{\bar{y}(i) - \mu(i)\}.$$

So an alternative expression for the log likelihood is

$$\ell = \sum_{i} n(i) \log p(i) - Nq \log(2\pi)/2 - \sum_{i} n(i) \log \det(\Sigma)/2$$
$$- N \operatorname{tr}(S \Sigma^{-1})/2 - \sum_{i} n(i) \{\bar{y}(i) - \mu(i)\}^{\top} \Sigma^{-1} \{\bar{y}(i) - \mu(i)\}/2.$$

The full homogeneous model has MLEs $\hat{p}(i) = n(i)/N$, (so that $\hat{m}(i) = N\hat{p}(i)$), $\hat{\mu}(i) = \bar{y}(i)$, and $\hat{\Sigma} = S = \sum_i n(i)S_i/N$, so the maximized log likelihood for this model is

$$\hat{\ell}_s = \sum_i n(i) \log\{n(i)/N\} - Nq \log(2\pi)/2 - N \log\det(S)/2 - Nq/2, \quad (5.10)$$

and the deviance of a homogeneous model $\mathcal{M}$ with MLEs $\hat{p}(i)$, $\hat{\mu}(i)$, and $\hat{\Sigma}$ with respect to the full homogeneous model simplifies to

$$D = 2\sum_i n(i) \log\{n(i)/\hat{m}(i)\} - N \log\det(S\hat{\Sigma}^{-1}) + N\{\text{tr}(S\hat{\Sigma}^{-1}) - q\}$$

$$+ \sum_i n(i)\{\bar{y}(i) - \hat{\mu}(i)\}^\top \hat{\Sigma}^{-1}\{\bar{y}(i) - \hat{\mu}(i)\}.$$

Note that in contrast to the models considered in Chap. 4, we do not necessarily have $\text{tr}(S\hat{\Sigma}^{-1}) = q$ so the term $N \log\det(S\hat{\Sigma}^{-1})$ does not disappear.

5.7.2 Dimension of MI-models

The dimension of a mixed interaction model may be simply calculated by adding the dimensions of the component models for $g(i)$ and each $h^u(i)$ to the number of free elements of the covariance matrix, and finally subtract one for the normalisation constant.

5.7.3 Inference

Under $\mathcal{M}$, the deviance D is asymptotically $\chi^2(k)$ where the degrees of freedom k is the difference in dimension (number of free parameters) between the saturated model and $\mathcal{M}$. Similarly, for two nested models $\mathcal{M}_1 \subseteq \mathcal{M}_2$, the deviance difference $D_1 - D_2$ is asymptotically $\chi^2(k)$ where the degrees of freedom k is the difference in dimension (number of free parameters) between the two models.

5.7.4 Likelihood Equations

Suppose we have a sample of N independent, identically distributed observations (i^ν, y^ν) for $\nu = 1 \ldots N$. Let $(n(i), t(i), \bar{y}(i))_{i \in \mathcal{I}}$ be the observed counts, variate totals and variate means, for cell i, and SS and S be the uncorrected sums of squares and sample covariance matrices, i.e.,

$$n(i) = \#\{\nu : i^\nu = i\},$$

$$t(i) = \sum_{v:i^v=i} y^v,$$

$$\overline{y}(i) = t(i)/n(i),$$

$$SS = \sum_{v} y^v (y^v)^\top,$$

$$SSD = \sum_{i \in \mathcal{I}} \sum_{v:i^v=i} \{y^v - \overline{y}(i)\}\{y^v - \overline{y}(i)\}^\top = SS - n(i) \sum_{i \in \mathcal{I}} \overline{y}(i)\{\overline{y}(i)\}^\top$$

$$S = SSD/N$$

For $a \subseteq \Delta$, we write the marginal cell corresponding to i as i_a and likewise for $b \subseteq \Gamma$, we write the subvector of y as y_b. Similarly, we write the marginal cell counts as $\{n(i_a)\}_{i_a \in \mathcal{I}_a}$, marginal variate totals as $\{t^b(i_a)\}_{i_a \in \mathcal{I}_a}$ and marginal variate means as $\{\overline{y}_b(i_a)\}_{i_a \in \mathcal{I}_a}$. Define

$$SSD_a^b(i_a) = \sum_{v:i_a^v=i_a} \{y_b^k - \overline{y}_b(i_a)\}\{y_b^k - \overline{y}_b(i_a)\}^\top$$

and let

$$SSD_a^b = \sum_{i_a \in \mathcal{I}_a} SSD_a^b(i_a) = SS^b - \sum_{i_a \in \mathcal{I}_a} n(i_a)\overline{y}_b(i_a)\overline{y}_b(i_a)^\top$$

where SS^b is the b-submatrix of the sums-of-squares matrix SS.

The log-likelihood for the sample is

$$l = \sum_{(a,b) \in \mathcal{C}} \sum_{i_a \in \mathcal{I}_a} n(i_a)g_a(i_a) + \sum_{(a,b) \in \mathcal{C}} \sum_{i_a \in \mathcal{I}_a} h_a^b(i_a)^\top t^b(i_a)$$

$$- \sum_{u \in \Gamma} SS^{uu}k_{uu}/2 - \sum_{\{u,v\} \in \Gamma} SS^{uv}k_{uv} \qquad (5.11)$$

where in the last term there is a contribution from SS^{uv} only if $k_{uv} \neq 0$, that is if $\{u, v\} \in b$ for some generator $(a, b) \in \mathcal{C}$.

Consider now a given model with generators $\mathcal{C} = \{G_1, \ldots, G_m\}$ and derive the formulae for $g(i)$ and each $h^u(i)$ as described in Sect. 5.5. Then a set of *minimal sufficient statistics* is given by

1. A set of marginal tables of cell counts $\{n(i_a)\}_{i_a \in \mathcal{I}_a}$ for each discrete generator a.
2. For each $u \in \Gamma$, a set of marginal variate totals $\{t^u(i_a)\}_{i_a \in \mathcal{I}_a}$ for each linear generator a of u.
3. A set of marginal tables of uncorrected sums and squares $\{SS^b\}$ for each quadratic generator b.

From exponential family theory, we know that the MLE of $\{p(i), \mu(i), \Sigma\}$ can be found by equating the expectations of these minimal sufficient statistics to their observed values. Equating the minimal sufficient statistics to their observed values for a generator (a, b) yields:

$$n(i_a) = Np(i_a), \quad \forall i_a \in \mathcal{I}_a, \tag{5.12}$$

$$t^b(i_a) = N \sum_{j:j_a=i_a} p(j)\mu^b(j), \quad \forall i_a \in \mathcal{I}_a \tag{5.13}$$

$$SS^b = N \left\{ \Sigma^b + \sum_{j \in \mathcal{I}} p(j)\mu^b(j)\mu^b(j)^\top \right\}. \tag{5.14}$$

Each generator $(a, b) \in \mathcal{C}$ defines a set of equations of the form (5.12)–(5.14) and the collection of these equations are the *likelihood equations* for the model. The MLEs, when they exist, are the unique solution to these equations that also satisfy the model constraints.

For example, for the saturated model on $V = \Delta \cup \Gamma$, we set $a = \Delta$ and $b = \Gamma$. Here there are no model constraints, and from the equations we find that the MLEs are given as $\hat{p}(i) = n(i)/N$, $\hat{\mu}(i) = \bar{y}(i)$ and $\hat{\Sigma} = S$.

5.7.5 *Iterative Proportional Scaling*

As with discrete log-linear models and graphical Gaussian models, iterative methods to find the maximum likelihood parameter estimates are generally necessary. The iterative proportional scaling algorithm for mixed interaction models proceeds by equating observed and expected margins, in much the same way as with discrete and continuous models. An important conceptual difference, however, relates to marginalization. Whereas multinomial and Gaussian distributions are preserved under marginalization, the same is not generally true in the mixed case: the marginal distribution of a CG-distribution is not necessarily CG. For this reason the concept of *weak marginals* is needed.

5.7.5.1 Weak Marginals

Consider a CG-density f_V defined over the variables $V = \Delta \cup \Gamma$. Letting $a \subset \Delta$ and $b \subset \Gamma$ we wish to obtain the marginal density $f_{a \cup b}$. This density is obtained by first integrating over $y_{\Gamma \setminus b}$ to produce $f_{\Delta \cup b}$ which again is a CG-density. The next step is to sum over $i_{\Delta \setminus a}$ to form $f_{a \cup b}$. This summation may involve forming a mixture of normal densities, which does not generally have the form of a CG-density. However, even though $f_{a \cup b}$ is not in general a CG-density we can find the moments of $f_{a \cup b}$ using standard formulae, namely

$$p_{[a]}(i_a) = p(I_a = i_a) = p(i_a) = \sum_{j:j_a=i_a} p(j)$$

$$\mu^b_{[a]}(i_a) = \mathrm{E}(Y^b \mid I_a = i_a) = \sum_{j:j_a=i_a} \frac{p(j)}{p_{[a]}(i_a)}\mu^b(j), \quad \text{and}$$

$$\Sigma_{[a]}^{b}(i_a) = \mathbf{V}(Y^b \mid I_a = i_a)$$

$$= \Sigma^b + \sum_{j:j_a=i_a} \frac{p(j)}{p_{[a]}(i_a)} \{\mu^b(j) - \mu_{[a]}^{b}(i_a)\}\{\mu^b(j) - \mu_{[a]}^{b}(i_a)\}^{\top}.$$

These moments $\{p_{[a]}(i_a), \mu_{[a]}^{b}(i_a), \Sigma_{[a]}^{b}(i_a)\}_{i_a \in \mathcal{I}_a}$ define a CG density $f_{[a \cup b]}$ denoted the *weak marginal density* (which is not homogeneous).

Furthermore, we define the *homogeneous weak marginal variance* to be:

$$\Sigma_{[a]}^{b} = \sum_{i_a \in \mathcal{I}_a} p_{[a]}(i_a) \Sigma_{[a]}^{b}(i_a)$$

$$= \Sigma^b + \sum_{i_a \in \mathcal{I}_a} \sum_{j:j_a=i_a} p(j)\{\mu^b(j) - \mu_{[a]}^{b}(i_a)\}\{\mu^b(j) - \mu_{[a]}^{b}(i_a)\}^{\top}.$$

The moments $\{p_{[a]}(i_a), \mu_{[a]}^{b}(i_a), \Sigma_{[a]}^{b}\}_{i_a \in \mathcal{I}_a}$ define a CG density $f_{[a \cup b]}^{h}$ which is denoted the *homogeneous weak marginal density*.

The weak marginal density is the CG-density which best approximates the true marginal $f_{a \cup b}$ in the sense of minimizing the Kullback–Leibler distance, see Lauritzen (1996), p. 162. The same proof yields that the analogous statement holds for the homogeneous weak marginal.

5.7.5.2 Likelihood Equations Revisited

It is illustrative to rewrite the likelihood equations as follows. Observe that

$$Q_a^b = \sum_{i_a \in \mathcal{I}_a} \sum_{j:j_a=i_a} p(j)\{\mu^b(j) - \mu_{[a]}^{b}(i_a)\}\{\mu^b(j) - \mu_{[a]}^{b}(i_a)\}^{\top}$$

$$= \sum_{i \in \mathcal{I}} p(i)\mu^b(i)\{\mu^b(i)\}^{\top} - \sum_{i_a \in \mathcal{I}_a} p_{[a]}(i_a)\mu_{[a]}^{b}(i_a)\{\mu_{[a]}^{b}(i_a)\}^{\top} \quad (5.15)$$

Using the definitions of the parameters of weak marginal models, (5.12) and (5.13) imply that

$$n(i_a)/N = p_{[a]}(i_a), \qquad \bar{y}^b(i_a) = t^b(i_a)/n(i_a) = \mu_{[a]}^{b}(i_a). \quad (5.16)$$

Using (5.15) and (5.16) we get from (5.14) that

$$SSD_a^b = SS^b - \sum_{i_a \in \mathcal{I}_a} n(i_a)\bar{y}^b(i_a)\bar{y}^b(i_a)^{\top}$$

$$= N\left[\Sigma^b + \sum_{i_a \in \mathcal{I}_a} \sum_{j:j_a=i_a} p(j)\{\mu^b(j) - \mu_{[a]}^{b}(i_a)\}\{\mu^b(j) - \mu_{[a]}^{b}(i_a)\}^{\top}\right]$$

$$= N(\Sigma^b + Q_a^b) = N\Sigma_{[a]}^{b}$$

The MLEs under the saturated MI-model for the variables $a \cup b$ (whose density is denoted $\tilde{f}_{a \cup b}$) are $\{\tilde{p}_a(i_a), \tilde{\mu}_a^b(i_a), \tilde{S}_a^b\}_{i_a \in \mathcal{I}_a}$ where

$$\tilde{p}_a(i_a) = n(i_a)/N, \qquad \tilde{\mu}_a^b(i_a) = \bar{y}^b(i_a) \quad \text{and} \quad \tilde{S}_a^b = SSD_a^b/N.$$

In other words, the likelihood equations are:

$$\tilde{p}_a(i_a) = n(i_a)/N = p_{[a]}(i_a) \tag{5.17}$$

$$\tilde{\mu}_a^b(i_a) = \bar{y}^b(i_a) = \mu_{[a]}^b(i_a) \tag{5.18}$$

$$\tilde{S}_a^b = SSD_a^b/N = \Sigma_{[a]}^b \tag{5.19}$$

thus the homogeneous weak marginal model on $a \cup b$ should be identical to the saturated MI-model on $a \cup b$, i.e. $f_{[a \cup b]}^h = \tilde{f}_{a \cup b}$.

5.7.5.3 General IPS Update Step

Here we describe the iterative algorithm for general MI-models implemented in **gRim** and MIM (Edwards 2000). Equations (5.17)–(5.19) suggest the following IPS update step for a generator (a, b):

$$f^*(i, y) \propto f(i, y) \frac{f_{a \cup b}^{sat}(i_a, y_b)}{f_{[a \cup b]}^h(i_a, y_b)} \tag{5.20}$$

Note that the right-hand side of (5.20) will not in general be a density: Integrating over $y_{\Gamma \setminus b}$ and summing over $i_{\Delta \setminus a}$ gives

$$f_{a \cup b}(i_a, y^b) f_{a \cup b}^{sat}(i_a, y^b)/f_{[a \cup b]}^h(i_a, y^b)$$

which will not be a density unless the marginal density $f_{a \cup b}(i_a, y_b)$ equals the homogeneous weak marginal density $f_{[a \cup b]}^h(i_a, y_b)$.

It is convenient to perform the update (5.20) on log-densities using the canonical parametrisation, since it just involves to addition and subtraction of canonical parameters. From (5.17)–(5.19), to update (g, h, K) we first transform the moment parameters $\{\tilde{p}_a, \tilde{\mu}_a^b, \tilde{S}_a^b\}$ and $\{p_{[a]}, \mu_{[a]}^b, \Sigma_{[a]}^b\}$ of $\tilde{f}_{a \cup b}$ and $f_{[a \cup b]}^h$ to canonical parameters $(\tilde{g}_a, \tilde{h}_a^b, \tilde{K}_a^b)$ and $(g_{[a]}, h_{[a]}^b, K_{[a]}^b)$. Then we

1. Update g as follows: For each $i_a \in \mathcal{I}_a$ do for all j for which $j_a = i_a$:

$$g(j) \leftarrow g(j) + \{\tilde{g}_a(i_a) - g_{[a]}(i_a)\}. \tag{5.21}$$

2. Update the b subvector of h as follows: For each $i_a \in \mathcal{I}_a$ do for all j for which $j_a = i_a$:

$$h^b(j) \leftarrow h^b(j) + \{\tilde{h}_a^b(i_a) - h_{[a]}^b(i_a)\}. \tag{5.22}$$

3. Update the b submatrix K^{bb} of K as follows:

$$K^{bb} \leftarrow K^{bb} + \{\tilde{K}_a^b - K_{[a]}^b\}. \tag{5.23}$$

After the update steps (5.21)–(5.23) we know h and K and hence the conditional distribution of y given i. To complete the update we must transform (g, h, K) to moment form (p, μ, Σ), normalize p to sum to one and transform back to canonical form (g, h, K) again before moving on to the next generator. Running through the generators $(a_1, b_1), (a_2, b_2), \ldots, (a_M, b_M)$ as described above constitutes one cycle of the iterative fitting process.

A measure of how much the updates (5.21)–(5.23) change the parameter estimates may be obtained by comparing the moments of $\tilde{f}_{a\cup b}$ and $f_{[a\cup b]}^h$. Following Edwards (2000) we use the quantity:

$$\texttt{mdiff}(a, b) = \max_{i_a \in \mathcal{I}_a, u, v \in b} \left\{ \frac{N|p_{[a]}(i_a) - \tilde{p}_a(i_a)|}{\sqrt{Np_{[a]}(i_a) + 1}}, \frac{|\mu_{[a]}^u(i_a) - \tilde{\mu}_a^u(i_a)|}{\sqrt{(\Sigma_{[a]}^b)_{uu}}}, \right.$$

$$\left. \frac{|(\Sigma_{[a]}^b)_{uv} - (\tilde{\Sigma}_a^b)_{uv}|}{\sqrt{(\Sigma_{[a]}^b)_{uu}(\Sigma_{[a]}^b)_{vv} + (\Sigma_{[a]}^b)_{uv}^2}} \right\} \tag{5.24}$$

It sometimes happens that the updates (5.21)–(5.23) lead to a decrease in the likelihood. To avoid this situation we first calculate $\texttt{mdiff}(a, b)$ in (5.24). If $\texttt{mdiff}(a, b)$ is smaller than some prespecified criterion we do not update the model but proceed to the next generator. If this is true for all generators we exit the iterative process, as it essentially only happens when we are close to the MLE.

Since the estimation algorithm in the $\texttt{mmod()}$ function is based on the model formula, which is not unique, there will be efficiency differences between the different representations of the same model. The maximal form is the most efficient.

5.7.5.4 Step-Halving Variant

It can happen that the updates (5.21)–(5.23) fail to increase the likelihood, or lead to a K that is not positive definite. The step-halving variant of the algorithm (currently not implemented in **gRim**) replaces the three update steps in (5.21)–(5.23) with:

$$g(j) \leftarrow g(j) + \kappa\{\tilde{g}_a(i_a) - g_{[a]}(i_a)\},$$

$$h^b(j) \leftarrow h^b(j) + \kappa\{\tilde{h}_a^b(i_a) - h_{[a]}^b(i_a)\},$$

$$K^{bb} \leftarrow K^{bb} + \kappa\{\tilde{K}_a^b - K_{[a]}^b\}.$$

Initially $\kappa = 1$. The update is attempted and it is then checked if (1) K is positive definite and (2) the likelihood is increased. If either of these conditions fail, κ is halved and the update is attempted again. The step-halving variant is therefore slower than the general algorithm. Edwards (2000, p. 312) shows an example with contrived data where step-halving is necessary.

5.7.5.5   Mixed Parameterisation Variant

If the model is collapsible onto the discrete parameters, the estimate $\hat{p}(i)$ is identical to the estimate obtained in the log-linear model with the same discrete generator. This permits another variant based on the mixed parametrisation to be used. It has the following update scheme

$$p(j) \leftarrow p(j)\{p(i_a)/p_{[a]}(i_a)\},$$

$$h^b(j) \leftarrow h^b(j) + \kappa\{\tilde{h}_a^b(i_a) - h_{[a]}^b(i_a)\},$$

$$K^{bb} \leftarrow K^{bb} + \kappa\{\tilde{K}_a^b - K_{[a]}^b\}.$$

The model is collapsible onto Δ if and only every connected component of the subgraph induced by the continuous variables has a complete boundary in the subgraph induced by the discrete variables (Frydenberg 1990b). This variant is currently not implemented in **gRim**.

5.8   Using gRim

The function `mmod()` in the **gRim** package allows homogeneous mixed interaction models to be defined and fitted to data.

```
> glist    <- ~treat:fat:protein+protein:lactose

~treat:fat:protein + protein:lactose

> milk <- mmod(glist, data=milkcomp1)

Model: A mModel with 4 variables
 graphical :  TRUE  decomposable :   TRUE
 -2logL    :        428.47 mdim :   26 aic :       480.47
 ideviance :         18.97 idf  :   14 bic :       532.66
 deviance  :          2.11 df   :    7
```

This model is shown in Fig. 5.1. More details about the model are obtained with

```
> summary(milk)

Mixed interaction model:
Generators:
  :"treat" "fat" "protein"
  :"protein" "lactose"
Discrete:   1   Continuous:    3
Is graphical: TRUE  Is decomposable: TRUE
logL: -214.233011, iDeviance: 241.774364
```

The parameters are obtained using `coef()` where the desired parameterization can be specified. For example, the canonical parameters are

```
> coef(milk, type="ghk")
```

```
MIparms: form=ghk
              a          b          c         d          e          f
[1,]   -676.055  -666.0859  -675.0546  -690.992  -664.9730  -666.7805
[2,]     -1.135    -0.2838    -0.9179    -1.022    -0.2012    -0.5375
[3,]     84.349    81.3414    83.8954    86.851    81.0040    81.8196
[4,]    164.634   164.6335   164.6335   164.634   164.6335   164.6335
              g
[1,]   -680.022       NA       NA       NA
[2,]     -1.043   0.5026   -0.815   0.000
[3,]     84.953  -0.8150   10.762   5.667
[4,]    164.634   0.0000    5.667  24.646
```

5.8.1 Updating Models

Models are changed using the update() method. A list with one or more of the
components add.edge, drop.edge, add.term and drop.term is specified. The
updates are made in the order given. For example:

```
> milk2 <- update(milk, list(add.edge=~fat:lactose,
                 drop.edge=~treat:protein))
```

```
Model: A mModel with 4 variables
 graphical :   TRUE  decomposable :   TRUE
 -2logL    :         446.17 mdim :   21 aic :        488.17
 ideviance :          10.12 idf  :    9 bic :        530.33
 deviance  :          10.96 df   :   12
```

5.8.2 Inference

Functions such as ciTest(), testInEdges(), testOutEdges(), etc. behave
more or less as for pure discrete and pure continuous variables. For example

```
> ciTest(milkcomp1)
```

```
Testing treat _|_ fat | protein dm lactose
Statistic (DEV):    4.371 df: 6 p-value: 0.6266 method: CHISQ
```

and

```
> testInEdges(milk,getInEdges(milk$glist))
```

```
  statistic df p.value     aic      V1      V2 action
1     5.530  6 0.47780  -6.470     fat   treat      +
2     9.345  6 0.15510  -2.655 protein   treat      +
3     4.139  1 0.04191   2.139 protein     fat      -
4     5.123  1 0.02362   3.123 lactose protein      -
```

```
> testOutEdges(milk,getOutEdges(milk$glist))
```

```
  statistic df p.value     aic      V1    V2 action
1    1.9464  6  0.9246  10.054 lactose treat      -
2    0.4914  1  0.4833   1.509 lactose   fat      -
```

or

```
> milk3 <- update(milk, list(drop.edge=~treat:protein))
```

```
Model: A mModel with 4 variables
 graphical :  TRUE  decomposable :  TRUE
 -2logL    :           447.16 mdim :   20 aic :      487.16
 ideviance :             9.63 idf  :    8 bic :      527.30
 deviance  :            11.45 df   :   13
```

```
> compareModels(milk, milk3)
```

```
Large:
  :"treat" "fat" "protein"
  :"protein" "lactose"
Small:
  :"protein" "lactose"
  :"treat" "fat"
  :"fat" "protein"
-2logL:    18.69 df: 6 AIC(k= 2.0):     6.69 p.value: 0.155100
```

and

```
> testdelete(milk, c("treat","protein"))
```

```
dev:    9.345 df:  6 p.value: 0.15510 AIC(k=2.0):   -2.7 edge:
        treat:protein
Notice: Test perfomed by comparing likelihood ratios
```

```
> testadd(milk, c("treat","lactose"))
```

```
dev:    1.946 df:  6 p.value: 0.92456 AIC(k=2.0):   10.1 edge:
        treat:lactose
Notice: Test perfomed by comparing likelihood ratios
```

5.8.3 Stepwise Model Selection

The stepwise() function in the **gRim** package implements stepwise selection for mixed interaction models. The functionality is very similar to that described above in Sect. 2.4 and Sect. 4.4.1, for discrete graphical models and undirected graphical Gaussian models respectively. We refer to those sections for further details, and illustrate using the wine dataset described in Sect. 5.2. We start from the saturated model and use the BIC criterion:

```
> data(wine, package=`gRbase')
> mm <- mmod(~.^., data=wine)
> mm2 <- stepwise(mm, k=log(nrow(wine)), details=0)
```

The selected model is shown below:

```
> plot(mm2)
```

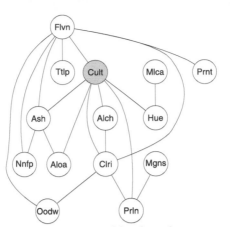

We note that the model is non-decomposable, since there are several chordless four-cycles in the graph. Since the graph is connected, it appears that all constituents differ over the grape varieties. Seven constituents are adjacent to the discrete variable. The model implies that these seven are sufficient to predict grape variety, since the remaining six are independent of variety given the seven, and so would not increase predictive ability.

5.9 An Example of Chain Graph Modelling

In this section we illustrate an approach that is appropriate when there is a clear overall response to the data, that is, when some variables are prior or explanatory to others, that are themselves prior or explanatory to others, and so on. The variables can *a priori* be divided into blocks, whose mutual ordering in this respect is clear. The goal of the analysis is to model the data, respecting this ordering between blocks, but not assuming any ordering within blocks. Chain graph models fit this purpose well.

The `Nutrimouse` dataset described above in Sect. 5.2 is here used as example. Here, the variables fall into three blocks: two discrete design variables (genotype and diet), 120 gene expression variables, and 21 lipid measurements. Clearly the design variables, which are subject to the control of the experimenter, are causally prior to the others. It is also natural as a preliminary working hypothesis to suppose that the gene expression measurements are causally prior to the lipid measurements, and this is the approach taken here. More advanced methods would be necessary to study whether there is evidence of influence in the opposite direction.

The chain graph is constructed using two graphical models: the first is for the gene expressions (block 2) given the design variables (block 1), and the second is for the lipids (block 3) given blocks 1 and 2. We use the **gRapHD** package described in Chap. 7. This package supports decomposable mixed models, both homogeneous and heterogeneous, exploiting the closed-form expressions for the MLEs (5.9). This restriction also means that models can simply be specified as graphs, rather than using model formulae.

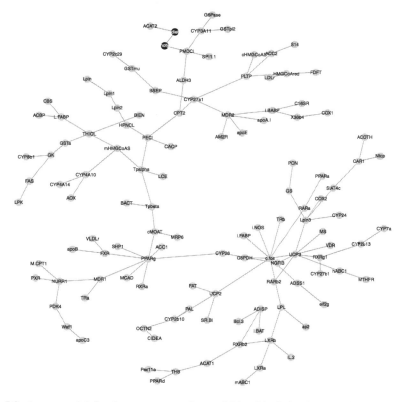

Fig. 5.3 A tree model for the gene expression variables (block 2) given the design variables (block 1)

To model the conditional distribution of block 2 given block 1 we restrict attention to models in which block 1 is complete, that is, there is an edge between the two design variables. See Fig. 4.33. The following code first finds the minimal BIC forest containing this edge, and then uses this as initial model in a forward selection process to find the minimal BIC decomposable model. This takes a few seconds.

```
> data(Nutrimouse, package='gRbase')
> library(gRapHD)
> block2 <- Nutrimouse[,1:122]
> gF1 <- minForest(block2, cond=list(1:2))
> gD1 <- stepw(gF1, data=block2)

> xyD1 <- plot(gD1, numIt=5000, disp=F)
> plot(gF1, numIt=0, coord=xyD1)

> plot(gD1, numIt=0, coord=xyD1)
```

We display the two graphs in Figs. 5.3 and 5.4, using the same vertex coordinates for clarity. The vertex coordinates are saved in a matrix xyD1.

We now turn to modelling the conditional distribution of block 3 variables given the prior blocks. We adopt the same approach as before, first finding a minimal BIC

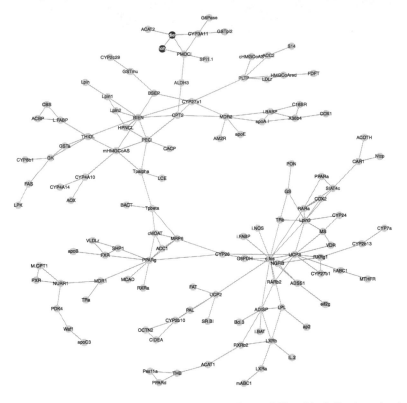

Fig. 5.4 A decomposable model for the gene expression variables (block 2) given the design variables (block 1)

forest and then using this as start model in a forward selection process. As before we restrict the search space to conditional models by including all edges between prior variables in the models considered. The forward selection process seeks the decomposable MI-model with minimum BIC in this search space.

```
> gF2 <- minForest(Nutrimouse, cond=list(1:122))
> gD2 <- stepw(gF2, data=Nutrimouse)
```

The stepw() function is computationally intensive, taking around 10 minutes on an ordinary laptop running Windows. We display the decomposable model in Fig. 5.5.

```
> plot(gD2, numIt=1000)
```

Now we construct a graph gD3 by adding to gD1 those edges in gD2 that have a vertex in block 3:

```
> E2 <- data.frame(gD2@edges)
> names(E2) <- c("v1", "v2")
> E2 <- as.matrix(E2[(E2$v1>122) | (E2$v2>122),])
> E3 <- rbind(gD1@edges, E2)
> gD3 <- gD2
> gD3@edges <- E3
```

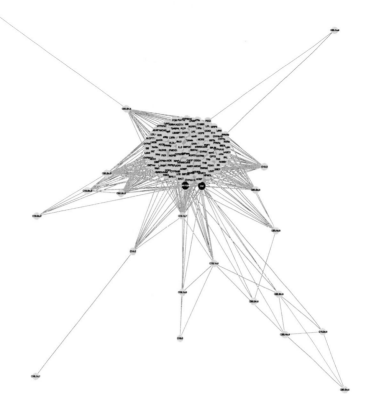

Fig. 5.5 A decomposable model for the lipid variables given the gene expression and design variables. The subgraph induced by the gene expression and design variables is complete, and is shown as a compact splat

Note that gD3 is an undirected graph rather than a chain graph. We use the **igraph** package to display it as a chain graph using different colours for the blocks, interblock edges displayed as arrows, in a layout in which the different blocks are separated for clarity. See Fig. 5.6

```
> # Define the blocks
> blk <- c(rep(1,2),rep(2,120),rep(3,21))
> # Derive the layout from the graph with only intrablock edges
> E <- gD3@edges
> E1 <- cbind(blk[E[,1]],blk[E[,2]])
> intrablock <- E1[,1]==E1[,2]
> tG3 <- gD3; tG3@edges <- E[intrablock,]
> itG3 <- as(as(tG3, "graphNEL"),"igraph")
> xy.coord <- piecewise.layout(itG3)
> # Use this for the chain graph
> igD3 <- graph.edgelist(E-1, directed=T)
> V(igD3)$label <- as.character(1:143)
> V(igD3)[blk==1]$color <- "white"
> V(igD3)[blk==2]$color <- "SkyBlue2"
```

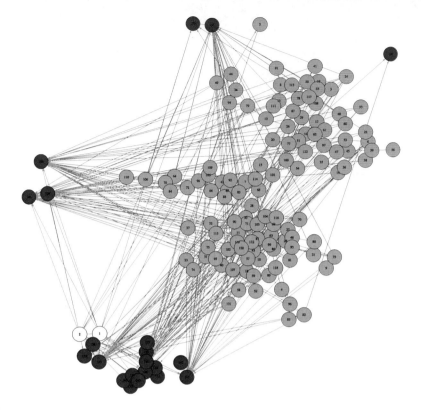

Fig. 5.6 A chain graph model for the nutrimouse data. The design variables are shown as *open circles*, the gene expression variables as *blue circles*, and the lipid variables as *red circles*. The variables are shown as column numbers

```
> V(igD3)[blk==3]$color <- "red"
> V(igD3)$size <- 8
> V(igD3)$label.cex <- 0.5
> E(igD3)[intrablock]$arrow.mode <- "-"
> E(igD3)[!intrablock]$arrow.mode <- "->"
> E(igD3)$arrow.size <-0.3
> plot(igD3, layout=xy.coord)
```

5.10 Various

Several other R packages are designed for graphical modelling with mixed discrete and Gaussian variables. The package **CoCo** (Badsberg 1991) implements undirected graphical (and hierarchical) models with mixed variables. The package **deal** (Bøttcher and Dethlefsen 2003) allows a Bayesian analysis using models for mixed variables based on DAGs, based on the conditional Gaussian distribution. Prior distributions for the model parameters are set and posterior distributions given data are derived. A heuristic search strategy for structural learning is also supported. The package **RHugin** also supports the use of Bayesian network models with mixed variables: see Chap. 3.

Chapter 6
Graphical Models for Complex Stochastic Systems

6.1 Introduction

In this chapter we describe the use of graphical models in a Bayesian setting, in which parameters are treated as random quantities on equal footing with the random variables. This allows complex stochastic systems to be modelled. This is one of the most successful application areas of graphical models; we give only a brief introduction here and refer to Albert (2009) for a more comprehensive exposition.

The paradigm used in Chaps. 2, 4 and 5 was that of identifying a joint distribution of a number of variables based on independent and identically distributed samples, with parameters unknown apart from restrictions determined by a log-linear, Gaussian, or mixed graphical model.

In contrast, Chap. 3 illustrated how a joint distribution for a Bayesian network may be constructed from a collection of conditional distributions; the network can subsequently be used to infer values of interesting unobserved quantities given evidence, i.e. observations of other quantitites. As parameters and random variables are on an equal footing in the Bayesian paradigm, we may think of the interesting unobserved quantitites as parameters and the evidence as data.

In the present chapter we follow this idea through in a general statistical setting. We focus mainly on constructing full joint distributions of a system of observed and unobserved random variables by specifying a collection of conditional distributions for a graphical model given as a directed acyclic graph with nodes representing all these quantities. Bayes' theorem is then invoked to perform the necessary inference.

6.2 Bayesian Graphical Models

6.2.1 Simple Repeated Sampling

In the simplest possible setting we specify the joint distribution of a parameter θ and data x through a *prior distribution* $\pi(\theta)$ for θ and a conditional distribution $p(x \mid \theta)$

S. Højsgaard et al., *Graphical Models with R*, Use R!,   145
DOI 10.1007/978-1-4614-2299-0_6, © Springer Science+Business Media, LLC 2012

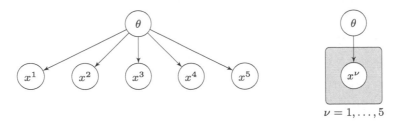

Fig. 6.1 Representation of a Bayesian model for simple sampling. The graph to the *left* indicates that observations are conditionally independent given θ; the picture to the *right* represents the same, but the plate allows a more compact representation

of data x for fixed value of θ, leading to the joint distribution

$$p(x, \theta) = p(x \mid \theta)\pi(\theta).$$

The prior distribution represents our knowledge (or rather uncertainty) about θ before the data have been observed. After observing that $X = x$ our *posterior distribution* $\pi^*(\theta)$ of θ is obtained by conditioning with the data x to obtain

$$\pi^*(\theta) = p(\theta \mid x) = \frac{p(x \mid \theta)\pi(\theta)}{p(x)} \propto L(\theta)\pi(\theta),$$

where $L(\theta) = p(x \mid \theta)$ is the *likelihood*. Thus *the posterior is proportional to the likelihood times the prior* and the normalizing constant is the marginal density $p(x) = \int p(x \mid \theta)\pi(\theta)d\theta$.

If the data is a sample $x = (x^1, x^2, x^3, x^4, x^5)$ we can represent this process by a small Bayesian network as shown to the left in Fig. 6.1. This network represents the model

$$p(x^1, \ldots, x^5, \theta) = \pi(\theta) \prod_{\nu=1}^{5} p(x^\nu \mid \theta).$$

reflecting that the individual observations are conditionally independent and identically distributed given θ. We can make a more compact representation of the network by introducing a plate which indicates repeated observations, such as shown to the right in Fig. 6.1.

For a more sophisticated example, consider a graphical Gaussian model given by the conditional independence $X_1 \perp\!\!\!\perp X_3 \mid X_2$ for fixed value of the concentration matrix K. In previous chapters we would have represented this model with its dependence graph:

However, in the Bayesian setting we need to include the parameters explicitly into the model, and could for example do that by the graph in Fig. 6.2.

The model is now represented by a chain graph, where the first chain component describes the structure of the prior distribution for the parameters in the concen-

Fig. 6.2 A chain graph representing N independent observations of $X = (X_1, X_2, X_3)$ from a Bayesian graphical Gaussian model in which $X_1^\nu \perp\!\!\!\perp X_3^\nu \mid X_2^\nu$, K and K follows a hyper Markov prior distribution

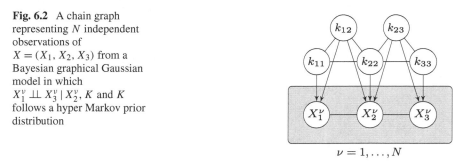

$$\nu = 1, \ldots, N$$

tration matrix. We have here assumed a so-called *hyper Markov prior distribution* (Dawid and Lauritzen 1993): conditionally on k_{22}, the parameters (k_{11}, k_{12}) are independent of (k_{23}, k_{33}). The plate indicates that there are N independent observations of X, so the graph has $3N + 5$ nodes. The chain component on the plate reflects the factorization

$$f(x_1, x_2, x_3 \mid K)$$
$$\propto \det(K)^{1/2} \exp\{-(x_1^2 k_{11} + x_2^2 k_{22} + x_3^2 k_{33} + 2x_1 x_2 k_{12} + 2x_2 x_3 k_{23})/2\}$$

for each of the individual observations of $X = (X_1, X_2, X_3)$.

6.2.2 Models Based on Directed Acyclic Graphs

A key feature of Bayesian graphical models is that explicitly including parameters and observations themselves in the graphical representation enables much more complex observational patterns to be accommodated. Consider for example a linear regression model

$$Y_i \sim N(\mu_i, \sigma^2) \quad \text{with } \mu_i = \alpha + \beta x_i \text{ for } i = 1, \ldots, N.$$

To obtain a full probabilistic model we must specify a joint distribution for (α, β, σ) whereas the dependent variables x_i are assumed known (observed). If we specify independent distributions for these quantities, Fig. 6.3 shows a plate- based representation of this model with α, β, and σ being marginally independent and independent of Y_i.

Note that μ_i are deterministic functions of their parents and the same model can also be represented without explicitly including these nodes. However, there can be specific advantages of representing the means directly in the graph. If the independent variables x_i are not centered, i.e. $\bar{x} \neq 0$, the model would change if x_i were replaced with $x_i - \bar{x}$, as α then would be the conditional mean when $x_i = \bar{x}$ rather than when $x_i = 0$, inducing a different distribution of μ_i.

For a full understanding of the variety and complexity of models that can easily be described by DAGs with plates, we refer to the manual for BUGS (Spiegelhalter et al. 2003), which also gives the following example.

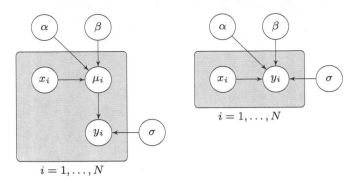

Fig. 6.3 Graphical representations of a traditional linear regression model with unknown intercept α, slope β, and variance σ^2. In the representation to the *left*, the means μ_i have been represented explicitly

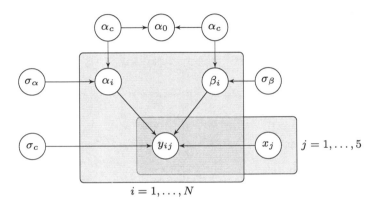

Fig. 6.4 Graphical representation of a random coefficient regression model for the growth of rats

Weights have been measured weekly for 30 young rats over five weeks. The observations Y_{ij} are the weights of rat i measured at age x_j. The model is essentially a random effects linear growth curve:

$$Y_{ij} \sim \mathcal{N}(\alpha_i + \beta_i(x_j - \bar{x}), \sigma_c^2)$$

and

$$\alpha_i \sim \mathcal{N}(\alpha_c, \sigma_\alpha^2), \qquad \beta_i \sim \mathcal{N}(\beta_c, \sigma_\beta^2),$$

where $\bar{x} = 22$. Interest particularly focuses on the intercept at zero time (birth), denoted $\alpha_0 = \alpha_c - \beta_c \bar{x}$. The graphical representation of this model is displayed in Fig. 6.4.

For a final illustration we consider the chest clinic example in Sect. 3.1.1. Figure 6.5 shows a directed acyclic graph with plates representing N samples from the chest clinic network.

Fig. 6.5 A graphical representation of N samples from the chest clinic network, with parameters unknown and marginally independent for seven of the nodes

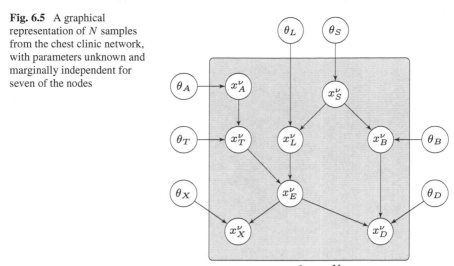

$$\nu = 1, \ldots, N$$

Here we have introduced a parameter node for each of the variables. Each of these nodes may contain parameters for the conditional distribution of a node given any configuration of its parents, so that, following Spiegelhalter and Lauritzen (1990), we would write for the joint model

$$p(x, \theta) = \prod_{v \in V} \pi(\theta_v) \prod_{\nu=1}^{N} p(x_v^\nu \mid x_{\text{pa}(v)}^\nu, \theta_v).$$

6.3 Inference Based on Probability Propagation

If the prior distributions of the unknown parameters are concentrated on a finite number of possibilities, i.e. the parameters are all discrete, the marginal posterior distribution of each of these parameters can simply be obtained by probability propagation in a Bayesian network with $7 + 8N$ nodes, inserting the observations as observed evidence. The moral graph of this network is shown in Fig. 6.6. This graph can be triangulated by just adding edges between x_L^ν and x_B^ν and the associated junction tree would thus have $10N$ cliques of size at most 4. Thus, propagation would be absolutely feasible, even for large N.

We illustrate this procedure in the simple case of $N = 3$ where we only introduce unknown parameters for the probability of visiting Asia and the probability of a smoker having lung cancer, each having three possible levels, low, medium and high. We first define the parameter nodes

```
> library(gRain)
> lmh <- c("low","medium","high")
> thA<- cptable(~theta_A, values =c(1,1,1), levels=lmh)
```

Fig. 6.6 Moral and triangulated graph of N samples from the chest clinic network, with seven unknown parameters

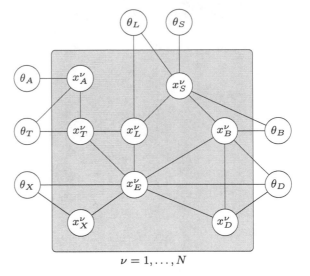

$$\nu = 1, \ldots, N$$

```
> thL<- cptable(~theta_L, values =c(1,1,1), levels=lmh)
> param <- list(thA, thL)
```

and then specify a template for probabilities where we notice that A and L have an extra parent

```
> yn <- c("yes","no")
> a     <- cptable(~asia[i]|theta_A, values=c(1,99,2,98,5,95),levels=yn)
> t.a  <- cptable(~tub[i]|asia[i], values=c(5,95,1,99),levels=yn)
> s    <- cptable(~smoke[i], values=c(5,5), levels=yn)
> l.s  <- cptable(~lung[i]|smoke[i]:theta_L,
+               values=c(5,95,1,99,1,9,1,99,1,4,1,99), levels=yn)
> b.s  <- cptable(~bronc[i]|smoke[i], values=c(6,4,3,7), levels=yn)
> e.lt <- cptable(~either[i]|lung[i]:tub[i],
+               values=c(1,0,1,0,1,0,0,1),levels=yn)
> x.e  <- cptable(~xray[i]|either[i], values=c(98,2,5,95), levels=yn)
> d.be <- cptable(~dysp[i]|bronc[i]:either[i],
+               values=c(9,1,7,3,8,2,1,9), levels=yn)
> plist.tmp <- list(a, t.a, s, l.s, b.s, e.lt, x.e, d.be)
```

We create three instances of the pattern defined above. In these instance the variable name `asia[i]` is replaced by `asia1`, `asia2` and `asia3` respectively.

```
> plate      <- repeatPattern(plist.tmp, instances=1:3)
```

We then proceed to the specification of the full network which is displayed in Fig. 6.7:

```
> plist    <- compileCPT(c(param, plate))
> chestlearn <-grain(plist)
> plot(chestlearn)
```

Finally we insert evidence for three observed cases, none of whom have been to Asia, all being smokers, one of them presenting with dyspnoea, one with a pos-

Fig. 6.7 Bayesian network
for the chest clinic example
with two unknown parameter
nodes and two potential
observations of the network.
Parameters appear as nodes in
the graph

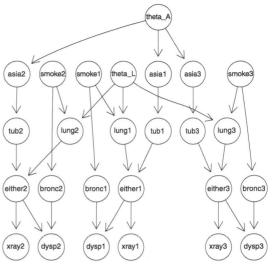

itive X-ray, one with dyspnoea and a negative X-ray; we then query the posterior
distribution of the parameters:

```
> chestlearn.ev<- setFinding(chestlearn,
+  nodes = c("asia1","smoke1","xray1"), c("no","yes","yes"))
> chestlearn.ev<- setFinding(chestlearn.ev,
+  nodes = c("asia2","smoke2","dysp2"), c("no","yes","yes"))
> chestlearn.ev<- setFinding(chestlearn.ev,
+  nodes = c("asia3","smoke3","dysp3","xray3"),
        c("no","yes","yes","no"))
> querygrain(chestlearn.ev,nodes =c("theta_A","theta_L"))

$theta_A
theta_A
   low medium   high
0.3504 0.3399 0.3096

$theta_L
theta_L
   low medium   high
0.2211 0.3099 0.4690
```

We see that the probabilities of visiting Asia is now more likely than before to be
low, whereas the probability of having lung cancer for a smoker is more likely to be
high.

In the special case where all cases have been completely observed, it is not necessary to form the full network with $7 + 8N$ nodes, but updating can be performed
sequentially as follows.

Let $p_n^*(\theta)$ denote the posterior distribution of θ given n observations $x^1, \ldots, x^n$, i.e. $p_n^*(\theta) = p(\theta \mid x^1, \ldots, x^n)$. We then have the recursion:

$$p_n^*(\theta) \propto p(x^1, \ldots, x^n, \theta) = \left\{ \prod_{\nu=1}^{n} p(x^\nu \mid \theta) \right\} p(\theta)$$

$$= p(x^n \mid \theta) \left\{ \prod_{\nu=1}^{n-1} p(x^\nu \mid \theta) \right\} p(\theta)$$

$$\propto p(x^n \mid \theta) p_{n-1}^*(\theta).$$

Hence we can incorporate evidence from the n-th observation by using the posterior distribution from the $n - 1$ first observations as a prior distribution for a network representing only a single case. It follows from the moral graph in Fig. 6.6 that if all nodes in the plates are observed, the seven parameters are conditionally independent also in the posterior distribution after n observations. If cases are incomplete, such a sequential scheme can only be used approximately (Spiegelhalter and Lauritzen 1990).

6.4 Computations Using Monte Carlo Methods

In most cases the posterior distribution

$$\pi^*(\theta) = p(\theta \mid x) = \frac{p(x \mid \theta) \pi(\theta)}{p(x)} \propto p(x \mid \theta) \pi(\theta) \tag{6.1}$$

of the parameters of interest cannot be calculated or represented in a simple fashion. This would for example be the case if the parameter nodes in Fig. 6.5 had values in a continuum and there were incomplete observations, such as in the example given in the previous section.

In such models one will often resort to Markov chain Monte Carlo (MCMC) methods: we cannot calculate $\pi^*(\theta)$ analytically but if we can generate samples $\theta^{(1)}, \ldots, \theta^{(M)}$ from the distribution $\pi^*(\theta)$, we can do just as well.

6.4.1 Metropolis–Hastings and the Gibbs Sampler

Such samples can be generated by the Metropolis–Hastings algorithm. In the following we change the notation slightly.

We suppose that we know $p(x)$ only up to a normalizing constant. That is to say, $p(x) = k(x)/c$, where $k(x)$ is known but c is unknown. We partition x into blocks, for example $x = (x_1, x_2, x_3)$.

We wish to generate samples $x^1, \ldots, x^M$ from $p(x)$. Suppose we have a sample $x^{t-1} = (x_1^{t-1}, x_2^{t-1}, x_3^{t-1})$ and also that x_1 has also been updated to x_1^t in the current iteration. The task is to update x_2. To do so we need to specify a proposal distribution h_2 from which we can sample candidate values for x_2. The single component Metropolis–Hastings algorithm works as follows:

1. Draw $x_2 \sim h_2(\cdot \,|\, x_1^t, x_2^{t-1}, x_3^{t-1})$. Draw $u \sim U(0, 1)$.
2. Calculate acceptance probability

$$\alpha = \min\left(1, \frac{p(x_2 \,|\, x_1^t, x_3^{t-1}) h_2(x_2^{t-1} \,|\, x_1^t, x_2, x_3^{t-1})}{p(x_2^{t-1} \,|\, x_1^t, x_3^{t-1}) h_2(x_2 \,|\, x_1^t, x_2^{t-1}, x_3^{t-1})} \right) \qquad (6.2)$$

3. If $u < \alpha$ set $x_2^t = x_2$; else set $x_2^t = x_2^{t-1}$.

The samples $x^1, \ldots, x^M$ generated this way will form an ergodic Markov chain that, under certain conditions, has $p(x)$ as its stationary distribution so that the expectation of any function of x can be calculated approximately as

$$\int f(x) p(x)\, dx = \lim_{M \to \infty} \frac{1}{M} \sum_{v=1}^M f(x^v) \approx \frac{1}{M} \sum_{v=1}^M f(x^v).$$

Note that $p(x_2 \,|\, x_1^t, x_3^{t-1}) \propto p(x_1^t, x_2, x_3^{t-1}) \propto k(x_1^t, x_2, x_3^{t-1})$ and therefore the acceptance probability can be calculated even though $p(x)$ may only be known up to proportionality.

A special case of the single component Metropolis–Hastings algorithm is the Gibbs sampler: If as proposal distribution h_2 we choose $p(x_2 \,|\, x_1^t, x_3^{t-1})$ then the acceptance probability becomes 1 because terms cancel in (6.2). The conditional distribution of a single component X_2 given all other components (X_1, X_3) is known as the full conditional distribution.

For a directed graphical model, the density of full conditional distributions can be easily identified:

$$f(x_i \,|\, x_{V \setminus i}) \propto \prod_{v \in V} f(x_v \,|\, x_{\mathrm{pa}(v)})$$

$$\propto f(x_i \,|\, x_{\mathrm{pa}(i)}) \prod_{v \in \mathrm{ch}(i)} f(x_v \,|\, x_{\mathrm{pa}(v)}) = f(x_i \,|\, x_{\mathrm{bl}(i)}), \qquad (6.3)$$

where $\mathrm{bl}(i)$ is the *Markov blanket* of node i:

$$\mathrm{bl}(i) = \mathrm{pa}(i) \cup \mathrm{ch}(i) \cup \left\{ \bigcup_{v \in \mathrm{ch}(i)} \mathrm{pa}(v) \setminus \{i\} \right\}$$

or, equivalently, the neighbours of i in the moral graph, see Sect. 1.4.1. Note that (6.3) holds even if some of the nodes involved in the expression correspond to values that have been observed. To sample from the posterior distribution of the unobserved values given the observed ones, only unobserved variables should be updated in the Gibbs sampling cycle.

In this way, a Markov chain of pseudo-observations from all unobserved variables is generated, and those corresponding to quantities (parameters) of interest can be monitored.

6.4.2 Using WinBUGS via R2WinBUGS

The program WinBUGS (Gilks et al. 1994) is based on the idea that the user specifies a Bayesian graphical model based on a DAG, including the conditional distribution of every node given its parents. WinBUGS then identifies the Markov blanket of every node and using properties of the full conditional distributions in (6.3), a sampler is automatically generated by the program. As the name suggests, Win-BUGS is available on Windows platforms only. WinBUGS can be interfaced from R via the **R2WinBUGS** package (Sturtz et al. 2005) and to do this, WinBUGS must be installed. **R2WinBUGS** works by calling WinBUGS, doing the computations there, shutting WinBUGS down and returning control to R.

The model described in Fig. 6.3 can be specified in the BUGS language as follows (notice that the dispersion of a normal distribution is parameterized in terms of the concentration τ where $\tau = \sigma^{-2}$):

```
model {
  for (i in 1:N) {
    Y[i] ~ dnorm(mu[i],tau)
    mu[i] <- alpha + beta*(x[i] - x.bar)
  }
  x.bar <- mean(x[])
  alpha ~ dnorm(0, 1.0E-6)
  beta  ~ dnorm(0, 1.0E-6)
  sigma ~ dunif(0,100)
  tau   <- 1/pow(sigma,2)
}
```

BUGS comes with a Windows interface in the program WinBUGS. To analyse this model in R we can use the package **R2WinBUGS**. First we save the model specification to a plain text file:

```
> cat(
+ "model {
+   for (i in 1:N) {
+     Y[i] ~ dnorm(mu[i],tau)
+     mu[i] <- alpha + beta*(x[i] - x.bar)
+   }
+   x.bar <- mean(x[])
+   alpha ~ dnorm(0, 1.0E-6)
+   beta  ~ dnorm(0, 1.0E-6)
+   sigma ~ dunif(0,100)
```

```
+    tau    <- 1/pow(sigma,2)
+  }",
+  file="linesModel.txt" )
```

We specify data:

```
> Y <- c(1,3,3,3,5)
> x <- c(1,2,3,4,5)
> N <- 5
```

As the sampler must start somewhere, we specify initial values for the unknowns:

```
> p.ini <- list(alpha = 0, beta = 0, sigma = 1)
```

We may now ask WinBUGS for a sample from the model:

```
> library(R2WinBUGS)
> lines.res <-
+   bugs(data = list( Y=Y, x=x, N=N ),
+        inits    = list( p.ini ),
+        param    = c("alpha","beta","sigma"),
+        model    = "linesModel.txt",
+        n.chains = 1,
+        ## Total number of samples, including burn-in:
+        n.iter   = 7000,
+        ## Burn-in values; will be discarded in subsequent analyses:
+        n.burnin = 5000,
+        ## Of the non-discarded samples only every 'n.thin'th
+            will be used.
+        n.thin   = 5,
+        bugs.directory = "c:/Programs/WinBUGS14/",
+        debug    = F,
+        clearWD  = TRUE )
```

The file lines.res contains the output. A simple summary of the samples is

```
> print(lines.res)

Inference for Bugs model at "linesModel.txt", fit using WinBUGS,
 1 chains, each with 7000 iterations (first 5000 discarded), n.thin = 5
 n.sims = 400 iterations saved
          mean  sd 2.5%  25%  50%  75% 97.5%
alpha     3.0 1.0  1.7  2.7  3.0  3.3   4.6
beta      0.9 0.7 -0.1  0.6  0.8  1.0   2.3
sigma     1.5 2.1  0.5  0.7  1.0  1.5   6.2
deviance 14.4 5.3  9.0 10.8 12.8 16.4  28.5

DIC info (using the rule, pD = Dbar-Dhat)
pD = 0.2 and DIC = 14.7
DIC is an estimate of expected predictive error (lower deviance is
                                                better).
```

We next convert the output to a format suitable for analysis with the **coda** package:

```
> library(coda)
> lines.coda <- as.mcmc.list(lines.res)
```

An summary of the posterior distribution of the monitored parameters is as follows:

```
> summary(lines.coda)
```

```
Iterations = 5001:6996
Thinning interval = 5
Number of chains = 1
Sample size per chain = 400
```

1. Empirical mean and standard deviation for each variable,
   plus standard error of the mean:

```
              Mean     SD Naive SE Time-series SE
alpha        2.980 1.037   0.0518         0.0525
beta         0.887 0.735   0.0367         0.0465
deviance    14.425 5.307   0.2654         0.3996
sigma        1.534 2.139   0.1070         0.1536
```

2. Quantiles for each variable:

```
              2.5%    25%    50%    75% 97.5%
alpha        1.708  2.708  3.023  3.28  4.56
beta        -0.065  0.598  0.813  1.04  2.35
deviance     9.046 10.837 12.775 16.41 28.47
sigma        0.459  0.740  1.002  1.49  6.16
```

As the observations are very informative, the posterior distributions of the regression parameters α and β are similar to the sampling distributions obtained from a standard linear regression analysis:

```
> summary(lm(Y~I(x-mean(x))))
```

```
Call:
lm(formula = Y ~ I(x - mean(x)))
```

```
Residuals:
        1         2         3         4         5
-4.00e-01  8.00e-01  4.84e-17 -8.00e-01  4.00e-01
```

```
Coefficients:
                Estimate Std. Error t value Pr(>|t|)
(Intercept)        3.000      0.327    9.19   0.0027 **
I(x - mean(x))     0.800      0.231    3.46   0.0405 *
---
Signif. codes:  0 '***' 0.001 '**' 0.01 '*' 0.05 '.' 0.1 ' ' 1
```

```
Residual standard error: 0.73 on 3 degrees of freedom
Multiple R-squared:  0.8,    Adjusted R-squared: 0.733
F-statistic:    12 on 1 and 3 DF,  p-value: 0.0405
```

A traceplot (see Fig. 6.8) of the samples is useful for visual inspection of indications that the sampler has not converged. There appears to be no problem here:

```
> library(coda)
> par(mfrow=c(2,2))
> traceplot(lines.coda)
```

A plot of the marginal posterior densities (see Fig. 6.9) provides a supplement to the numeric summaries shown above:

```
> par(mfrow=c(2,2))
> densplot(lines.coda)
```

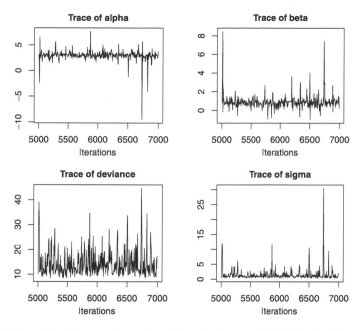

Fig. 6.8 A traceplot of the samples produced by BUGS is a useful tool for visual inspection of indications of that the sampler has not converged

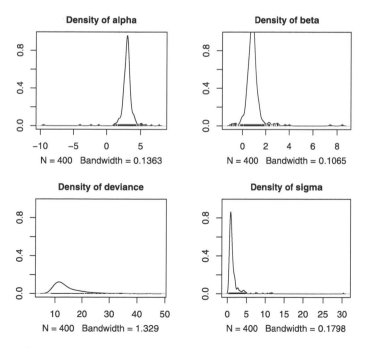

Fig. 6.9 A plot of each posterior marginal distribution provides a provides a supplement to the numeric summary statistics

6.5 Various

An alternative to WinBUGS is OpenBUGS (Spiegelhalter et al. 2011). The two programs have the same genesis and the model specification languages are very similar. OpenBUGS can be interfaced from R via the **BRugs** package and OpenBUGS/**BRugs** is available for all platforms. The modus operandi of **BRugs** is fundamentally different from that of WinBUGS: a sampler created using **BRugs** remains alive in the sense that one may call the sampler repeatedly from within R. Yet another alternative is package **rjags** which interfaces the JAGS program; this must be installed separately and is available for all platforms.

Chapter 7
High Dimensional Modelling

7.1 Introduction

This chapter describes and compares some methods available in R for selecting and working with high-dimensional graphical models. By 'high-dimensional' we are thinking of models with hundreds to tens of thousands of variables. Modelling such data has become of central importance in molecular biology and other fields, but is challenging. Many graph-theoretic operations scale poorly: for example, finding the cliques of a general undirected graph is known to be NP-hard. Model selection algorithms that work well in low dimensional applications may be quite infeasible for high dimensional ones. There can be statistical as well as algorithmic limitations: for example, for high-dimensional Gaussian data with modest numbers of observations, maximum likelihood estimates will not exist for complex models. Generally it is necessary to assume that relatively simple models are adequate to model high-dimensional data.

In Sect. 7.2 two example datasets are described. In Sect. 7.3 some model selection algorithms available in R are compared in respect to their scalability. Sections 7.4, 7.5 and 7.6 describe the use of some of the more scalable methods in more detail. Finally, in Sect. 7.7 we describe a Bayesian approach, showing how to identify the MAP (maximum a posteriori) forest for high-dimensional discrete data.

7.2 Two Datasets

We illustrate the methods in this chapter using two datasets. The first is supplied along with **gRbase** and is taken from a study comparing gene expression profiles in tumours taken from two groups of breast cancer patient, namely those with and those without a mutation in the p53 tumour suppression gene. See Miller et al. (2005) for a further description of the study.

```
> data(breastcancer)
> dim(breastcancer)
```

S. Højsgaard et al., *Graphical Models with R*, Use R!,
DOI 10.1007/978-1-4614-2299-0_7, © Springer Science+Business Media, LLC 2012

```
[1]    250 1001
> table(sapply(breastcancer, class))

factor numeric
     1    1000
> table(breastcancer$code)

  case control
    58     192
```

There are $N = 250$ observations and 1001 variables, comprising 1000 continuous variables (the log-transformed gene expression values) and one binary factor, code. There are 58 cases (with a p53 mutation) and 192 controls (without the mutation).

The second dataset comes from a large multinational project to study human genetic variation, the HapMap project (http://www.hapmap.org/). The dataset concerns a sample of 90 Utah residents with northern and western European ancestry, the so-called CEU population, and contains information on genetic variants and gene expression values for this sample. The genetic variants are SNPs (single nucleotide polymorphisms), that is to say, individual bases at particular loci on the genome that show variation in the population. In statistical terms SNPs are categorical variables with three levels—two homozygotes and a heterozygote. Around 10 million SNPs have been identified in humans. Datasets containing both SNP and gene expression data enable study of the the genetic basis for differences in gene expression. The data from this sample are supplied along with the package **GGtools** in the BioConductor repository. The code

```
> data(hmceuB36.2021)
> k <- 200
> ggdata <- data.frame(hmceuB36.2021$male,
+   as(smList(MAFfilter(hmceuB36.2021, lower=.1))
+   [["21"]][,1:k], "character"), t(exprs(hmceuB36.2021))[,1:k])
> ggdata[,1:(k+1)] <- lapply(ggdata[,1:(k+1)], factor)
```

loads an object hmceuB36.2021 containing SNP data from chromosomes 20 and 21, gene expression data and other phenotypic information recorded for individuals in the sample. In all it contains data on 199921 SNPs on chromosome 20, 50165 on chromosome 21, and expression values for 47293 genes. The above code fragment creates a dataframe ggdata by extracting the individuals sex, the first 200 SNPs from chromosome 21 and the first 200 log-transformed gene expression values from hmceuB36.2021. Prior to extraction the SNPs are filtered so that SNPs with a minimum allele frequency of less than 10% are discarded. Some values of the SNPs are missing, but here the missing values are coded as a distinct character value, so the SNPs are factors with up to four levels. The last line converts the discrete variables into factors.

7.3 Computational Efficiency

A thorough study of the computational efficiency of the algorithms described in this book would be a huge and complex task. In this section, we report on some timings

Model selection algorithms

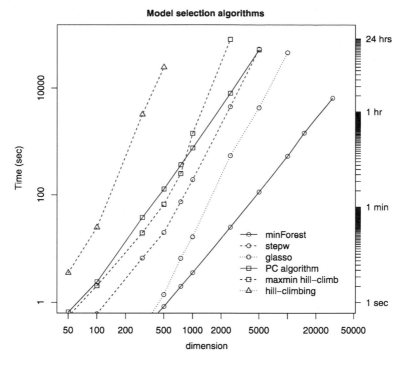

Fig. 7.1 Timing comparisons

of the algorithms when applied to a specific dataset, in a specific computing environment. It is hoped that this will give at least a rough impression of their relative efficiency on similar datasets in other computing environments.

The algorithms were applied to data from the HapMap project described in Sect. 7.2. For various values of the dimension p, the first p gene expression values were used. So there were 90 cases and p Gaussian variables, where p ranges from 50 to 50000. Three algorithms to select undirected graphical models, and four to select (equivalence classes of) graphical models based on DAGs were compared. The computations were run under Redhat Fedora 10 Linux on a Intel i7 four-core 2.93 GHz machine with 48 GB RAM. The timings are shown in Fig. 7.1.

The undirected model selection methods were:

(i) The extended Chow-Liu algorithm, implemented in the `minForest()` function in the **gRapHD** package, that finds the minimum BIC forest. This is further described in Sect. 7.4.

(ii) A greedy decomposable search algorithm, implemented in the `stepw()` function in the **gRApHD** package, that seeks (but is not guaranteed to find) the minimum BIC decomposable model. See Sect. 7.5 below. Here the minimum BIC forest is used as initial model.

(iii) The `glasso()` function in the **glasso** package described in Sect. 4.4.2. Here the tuning parameter $\rho = 0.2$ is used.

The DAG selection methods were:

(iv) The PC-algorithm implemented in the pc() function in the **pcalg** package, as described in Sect. 4.6.1. Here $\alpha = 0.05$ is used.
 (v) The hill-climbing algorithm implemented in the hc() function in the **bnlearn** package, as described in Sect. 4.6.2.1.
(vi) The max-min hill-climbing algorithm implemented in the mmhc() function in the **bnlearn** package, a hybrid constraint- and score-based algorithm, described above in Sect. 4.6.2.2. Here $\alpha = 0.05$ is used.

Note that (i) and (ii) return decomposable models, which represent equivalence classes of DAGs (see Sect. 4.5.1), so these can also be regarded as DAG selection methods.

We note that the algorithms for undirected models are more efficient than those for directed models. The most efficient of the latter is the pc-algorithm: however, when $p = 5000$, this takes approximately 24 hours whereas the extended Chow–Liu algorithm takes about 1 minute for these data.

7.4 The Extended Chow–Liu Algorithm

In a paper predating much of the theoretical development of graphical models, Chow and Liu (1968) described an algorithm to find the maximum likelihood tree model for multivariate discrete data. In modern terminology, tree models are discrete graphical models whose graphs are trees. Trees and forests are special cases of undirected graphs. A forest is an acyclic undirected graph, that is, an undirected graph with no cycles. A tree is a connected acyclic undirected graph. So a forest may have several connected components, these being trees. Chow and Liu showed that finding the maximum likelihood tree can be formulated as finding a maximum weight spanning tree—a task for which highly efficient algorithms exist. Their approach requires, first, that all edge weights are calculated, and then a maximum weight spanning tree algorithm is applied to find a maximum weight spanning tree. (This may be non-unique if there are ties in the edge weights.)

Usually the algorithm due to Kruskal (1956) is used to find the maximum weight spanning tree. This starts with the null graph and successively selects edges $e_1, \ldots, e_r$. If edges $e_1, \ldots, e_k$ have been selected, the algorithm selects an edge e such that

(a) $e \notin \{e_1, \ldots e_k\}$ and $\{e_1, \ldots e_k, e\}$ is a forest, and
(b) e has maximum weight among all edges satisfying (a).

Chow and Liu's approach may be extended in various ways (Edwards et al. 2010):

• It can be applied to Gaussian data using appropriate weights.
• By modifying the weights appropriately it can be adapted to find the minimal AIC or BIC forest. If this has several connected components we can analyze these separately—a dimension reduction that can be very useful with high-dimensional problems.

- It can be applied to mixed discrete and Gaussian data by modifying the weights appropriately and limiting the search space in (a) to strongly decomposable forests, that is, forests containing no forbidden paths. Recall that a forbidden path is a path between non-adjacent discrete nodes passing through continuous nodes. This restriction implies that in each tree of the forest, the discrete nodes induce a connected subgraph.
- In the conditional Chow–Liu algorithm (Kirshner et al. 2004) the search space is extended to graphs that include a given set of edges, E_0 say. Formally, the search space becomes

$$\{\mathcal{G} = (V, E) : E_0 \subseteq E \wedge \text{any cycle in } \mathcal{G} \text{ has all edges in } E_0\}$$

To do this, the algorithm starts off from $\mathcal{G}_0 = (V, E_0)$ and (a) is modified to restrict candidate edges to those that do not create *new* cycles.

The extended algorithm is implemented in the `minForest()` function in the **gRapHD** package. It requires the data to be supplied as a dataframe with discrete and/or continuous variables. The discrete variables must be represented as factors. For example, we can apply it to the `breastcancer` dataframe as follows:

```
> bF <- minForest(breastcancer)
> bF
```

```
gRapHD object
Number of edges       = 1000
Number of vertices    = 1001
Model                 = mixed and homogeneous
Statistic (minForest) = BIC
Statistic (stepw)     =
Statistic (user def.) =
Edges (minForest)     = 1...1000
Edges (stepw)         = 0...0
Edges (user def.)     = 1...1000
```

Per default, the `minForest` function returns the minimal BIC forest, in the form of a gRapHD object. Note that bF has 1001 nodes and 1000 edges. Since a forest with n nodes and k connected components has $n - k$ edges, we see that bF is a tree: all nodes are interconnected.

These gRapHD objects are essentially undirected graphs represented in node and edge list form, in which nodes are identified by their column numbers in the input dataframe. They also contain information on variable types (discrete or continuous) and names (which are used to label the nodes in plots). They may be displayed using the `plot` function

```
> plot(bF)
```

but here, plotting a high-dimensional graph like bF would not be a good idea: no structure would be visible. Instead, since we are primarily interested in the effect of the mutation on gene expression, let us look at the neighbourhood of the discrete variable, `code`. This is the last column of `breastcancer`, column number 1001. The following two lines of code extract the nodes of bF whose path length from `code` is less than or equal to 4, and then display the subgraph of bF induced by these nodes.

```
> nby <- neighbourhood(bF, orig=1001, rad=4)$v[,1]
> plot(bF, vert=nby, numIter=1000)
```

The `plot()` function, when applied to `gRapHD` objects, shows discrete variables as dots and continuous variables as circles. It uses the iterative layout algorithm of Fruchterman and Reingold (1991): here we specify 1000 iterations to get a clear layout. We see that the effect of the mutation on gene expression appears to be mediated by its effect on the expression of gene A.202870_s_at. To find out more about this gene we can google this string, from which we learn that under the alias CDC20 the gene ". . . appears to act as a regulatory protein interacting with several other proteins at multiple points in the cell cycle. It is required for two microtubule-dependent processes, nuclear movement prior to anaphase and chromosome separation." In other words, it is involved in cell division. Below, using strongly decomposable models, we re-examine the hypothesis that the effect of p53 mutation on gene expression is mediated by its effect on the expression of this gene.

The following code illustrates the extended Chow-Liu approach applied to the `ggdata` dataset. The minimal BIC forest is obtained using the `minForest()` function:

```
> ggF <- minForest(ggdata)
> ggF
```

```
gRapHD object
Number of edges        = 392
Number of vertices     = 401
Model                  = mixed and homogeneous
Statistic (minForest)  = BIC
Statistic (stepw)      =
Statistic (user def.)  =
Edges (minForest)      = 1...392
Edges (stepw)          = 0...0
Edges (user def.)      = 1...392

> table(Degree(ggF))

  0   1   2   3   4   5   6   7   8  10
  5 198 103  49  23   7  10   4   1   1

> plot(ggF, numIter=500, vert.labels=1:ggF@p, main="min BIC forest")
```

min BIC forest

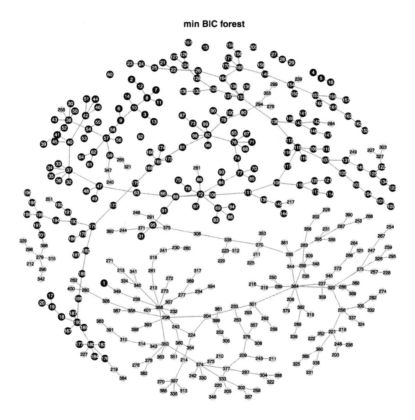

We see that ggF is a forest with $401 - 392 = 9$ connected components. The De-gree() function returns a vector containing the degree (number of adjacent nodes) of each node: we see that there are 5 isolated nodes. We can identify the components by converting the gRapHD object to a graphNEL object using the as() function, and then applying the connComp() function, which returns a list of components.

```
> cc <- connComp(as(ggF, "graphNEL"))
> sapply(cc, length)
```

```
[1] 172 218   3   1   3   1   1   1   1
```

The 9 connected components consist of two large components (with 172 and 218 nodes), two components with 3 nodes, and 5 isolated nodes. If we look at the two largest components

```
> intersect(cc[[1]], names(ggdata)[1:201])
```

```
[1] "hmceuB36.2021.male"
```

```
> length(intersect(cc[[2]], names(ggdata)[1:201]))
```

```
[1] 189
```

we see that the first contains only one discrete variable (sex), but the second contains 189 SNPs and 29 gene expression variables.

7.5 Decomposable Stepwise Search

In a celebrated paper, Chickering (1996) showed that identifying the Bayesian network that maximizes a score function is in general a NP-hard problem, and it is reasonable to suppose that this is also true of undirected graphical models (Markov networks). However, there are ways to improve computational efficiency. A useful approach is to restrict the search space to models with explicit estimates, the decomposable models. The following key result is exploited: if $\mathcal{M}_0 \subset \mathcal{M}_1$ are decomposable models differing by one edge $e = \{u, v\}$ only, then e is contained in one clique C of $\mathcal{M}_1$ only, and the likelihood ratio test for $\mathcal{M}_0$ versus $\mathcal{M}_1$ can be performed as a test of $u \perp\!\!\!\perp v | C \setminus \{u, v\}$. These computations only involve the variables in C. It follows that for likelihood-based scores such as AIC or BIC, score differences can be calculated locally—which is far more efficient then fitting both $\mathcal{M}_0$ and $\mathcal{M}_1$—and then stored, indexed by u, v and C, so that they can be reused again if needed in the course of the search. This can lead to considerable efficiency gains.

The stepw() function in the **gRapHD** package implements forward search through decomposable models to minimize the AIC or BIC. At each step, the edge giving the greatest reduction in AIC or BIC is added. A convenient choice of start model is the minimal AIC/BIC forest, but an arbitrary decomposable start model may be used. We illustrate use of this function by resuming the analysis of the breast cancer dataset. The minimal BIC forest for the neighbourhood of the code variable is obtained as follows.

```
> bc.marg <- breastcancer[,nby]
> mbF <- minForest(bc.marg)
> plot(mbF, numIter=1000)
```

The gene adjacent to code is A.202870_s_at (CDC20) as we saw before. This suggests that the effect of p53 mutation on gene expression is mediated by its effect on CDC20. However, this might be a consequence of adopting this restrictive—and sparse—model class. It is interesting to expand the search space to decomposable models. The minimal BIC decomposable model is obtained using the stepw() function:

```
> mbG <- stepw(model=mbF, data= bc.marg)
> mbG

gRapHD object
Number of edges      = 225
Number of vertices   = 94
Model                = mixed and homogeneous
Statistic (minForest) = BIC
Statistic (stepw)     = BIC
Statistic (user def.) =
Edges (minForest)    = 1...93
Edges (stepw)        = 94...225
Edges (user def.)    = 1...93
```

To plot mbG using the same layout as in the previous plot, we store the node coordinates from the previous plot and reuse them when plotting mbG, as follows:

```
> posn <- plot(mbF, numIter=1000, disp=F)
> plot(mbG, numIter=0, coord=posn)
```

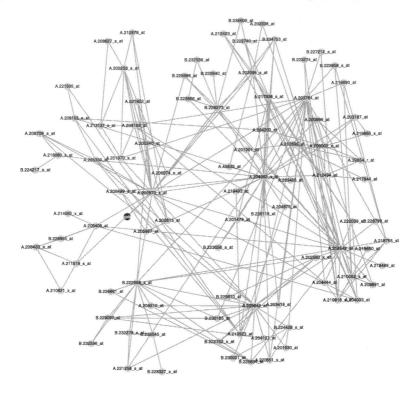

Although the minimal BIC decomposable model is considerably less sparse, the interpretation is unaltered: it still appears that the effect of p53 mutation on gene expression is mediated by its effect on the expression of CDC20.

An interesting aspect of this example is the presence of so-called *hub* genes—nodes of high degree—that may play a key role in the regulatory network. If we compare the degree distributions of the two graphs

```
> table(Degree(mbF))

 1  2  3  4  5  6 12 17
68 12  5  1  2  2  3  1

> table(Degree(mbG))

 1  2  3  4  5  6  7  8  9 12 15 21 22 25 29
 7 28 21 10  8  1  8  3  1  1  1  1  1  2  1

> Degree(mbF)[Degree(mbF)>4]

 2  6 13 18 20 24 27 31
12 17  6  6  5 12 12  5

> Degree(mbG)[Degree(mbF)>4]

 2  6 13 18 20 24 27 31
25 29 15 12 21 22 25  9
```

we see that the hub genes in the—presumably more realistic—graph mbG are reliably identified using the forest mbF.

7.6 Selection by Approximation

Here we illustrate use of the graphical lasso algorithm of Friedman et al. (2008)
described in Sect. 4.4.2. We apply it to the breastcancer dataset (omitting the discrete
class variable since the algorithm is only applicable to Gaussian data).

```
> S <- cor(breastcancer[,nby[-1]])
> res.lasso <- glasso(S, rho=0.8)
> AM <- res.lasso$wi != 0
> diag(AM) <- F
> rownames(AM) <- colnames(AM) <- names(breastcancer)[nby[-1]]
> g.lasso <- as(AM, "graphNEL")
> g.lasso

A graphNEL graph with undirected edges
Number of Nodes = 93
Number of Edges = 198

> g.HD <- as(g.lasso, "gRapHD")
> plot(g.HD, numIt=1000)
```

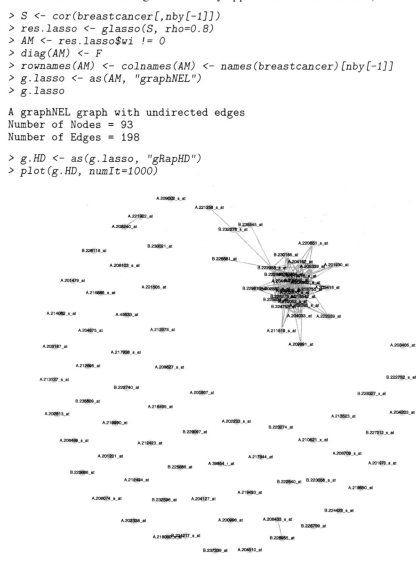

The example selects a model to the variables in the neighbourhood of the `code`
variable in the breast cancer dataset (omitting the `code` variable itself since it is discrete). We apply the `glasso()` function to the expirical correlation matrix of the
variables, in effect standardizing the variables to unit variance. Note that since the
glasso procedure is not scale invariant, this is normally a sensible step. As penalty

parameter we use $\rho = 0.8$: this choice was made so as to obtain a graph of comparable density to those obtained previously. The glasso() function returns a list containing the estimated inverse covariance matrix wi. Note that the method combines model selection with parameter estimation. In the code fragment shown we derive the adjacency matrix from the inverse covariance and use this to construct a graphNEL graph of the model. The diagonal elements of the adjacency matrix are set to false to omit self-edges from the graph.

The graph selected by the algorithm contains a module of interconnected variables and a large number of isolated ones. To see the former more closely, we can use the following code:

```
> cc <- connComp(g.lasso)
> sapply(cc, length)

 [1] 32  1  1  2  1  1  1  1  1  1  1  1  1  1  1  1  1  1  1  1  1  2
[23]  1  2  1  1  1  1  1  1  1  1  1  1  1  1  1  1  1  1  1  1  1  1
[45]  1  1  1  1  1  1  1  1  1  1  1  1  1  1  1

> g.sub <- subGraph(cc[[1]], g.lasso)
> g.subHD <- as(g.sub, "gRapHD")
> plot(g.subHD, numIt=1000)
```

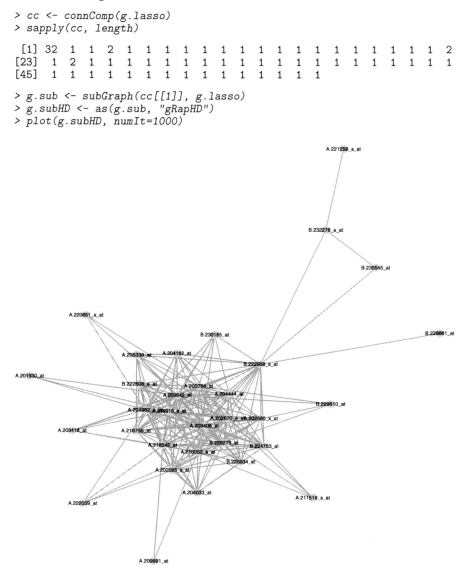

7.7 Finding MAP Forests

In this section we describe a Bayesian equivalent to the minimal AIC/BIC forest approach described in Sect. 7.4. This builds on the framework of Dawid and Lauritzen (1993). We write a collection of d discrete random variables as $X = (X_v)_{v \in V}$, and we write a generic observation as $i = (i_1, \ldots, i_d)$, and the observed data as $\mathcal{X} = (i^v, v = 1, \ldots, N)$. We are interested in a collection of graphs Γ with vertex set V. For each graph $\mathcal{G} \in \Gamma$, let $\Theta_{\mathcal{G}}$ be the associated parameter space, and $\mathcal{L}_{\mathcal{G}}$ be a prior distribution (or law) on $\Theta_{\mathcal{G}}$. Then the marginal likelihood of $\mathcal{G}$ is

$$f(\mathcal{X}|\mathcal{G}) = \int_{\Theta_{\mathcal{G}}} f(\mathcal{X}|\mathcal{G}, \theta) \mathcal{L}_{\mathcal{G}}(d\theta).$$

If $\pi(\mathcal{G})$ is the prior probability of $\mathcal{G}$, then the posterior probability is given as

$$\pi^*(\mathcal{G}) = \pi(\mathcal{G}|\mathcal{X}) \propto f(\mathcal{X}|\mathcal{G})\pi(\mathcal{G}). \tag{7.1}$$

The maximum a posteriori (MAP) estimate is the graph in Γ that maximizes $\pi^*(\mathcal{G})$.

We now sketch how a prior distribution $\mathcal{L}_{\mathcal{G}}$ on $\Theta_{\mathcal{G}}$ may be chosen. Consider first an unconstrained multinomial distribution on an array with parameters $p = (p(i))_{i \in \mathcal{I}}$, and let $\lambda = (\lambda(i))_{i \in \mathcal{I}}$ be an array of positive numbers. The Dirichlet distribution $D(\lambda)$ has density

$$\pi(p|\lambda) \propto \prod p(i)^{\lambda(i)-1}.$$

If the prior distribution of p is $D(\lambda)$ and counts $n = (n(i))_{i \in \mathcal{I}}$ are observed, then the posterior distribution of p is $D(\lambda + n)$: in other words, the Dirichlet distribution is the conjugate prior of the multinomial. The numbers λ are called the equivalent sample size, or smoothing parameter.

Dawid and Lauritzen (1993) generalize this to construct the conjugate prior for a decomposable graphical model $\mathcal{G}$, which they term the *hyper-Dirichlet* distribution. Essentially this involves specifying a Dirichlet prior for each clique of $\mathcal{G}$. Let $\mathcal{C} = (C_1, \ldots, C_k)$ be these cliques. Thus a hyper-Dirichlet prior is specified though the collection of arrays $(\lambda_C)_{C \in \mathcal{C}}$. These must satisfy a consistency criterion, namely that for all cliques $C, D \in \mathcal{C}$, $\lambda_C(i_{C \cap D}) = \lambda_D(i_{C \cap D})$ for all cells $i_{C \cap D}$. Without loss of generality we can specify the λ_C's by specifying a $\lambda = (\lambda(i))_{i \in \mathcal{I}}$ for the whole array and setting λ_C to the marginal totals $\lambda_C(i_C) = \sum_{j \in \mathcal{I}: j_C = i_C} \lambda(j)$. This construction automatically fulfills the consistency criteria. It also allows the array λ to function as a 'master-prior' to specify the smoothing parameters for the hyper-Dirichlet prior for the parameters for *any* decomposable model.

Dawid and Lauritzen (1993) also show that for a hyper-Dirichlet prior, the marginal likelihood factorizes in a fashion similar to the likelihood:

$$f(\mathcal{X}|\mathcal{G}) = \prod_{i=1...k} \frac{f(\mathcal{X}_{C_i}|\mathcal{G})}{f(\mathcal{X}_{S_i}|\mathcal{G})} \qquad (7.2)$$

where $\mathcal{S} = (S_1, \ldots, S_k)$ are the separators corresponding to $\mathcal{C}$. Moreover the factors $f(\mathcal{X}_{C_i}|\mathcal{G})$ are constant for all $\mathcal{G}$ in which C_i is (or is contained in) a clique, so in that sense the conditioning on $\mathcal{G}$ is unnecessary.

Let now Γ be the set of forests with vertex set V. These models are decomposable, and so using (7.2) we obtain for $\mathcal{G} \in \Gamma$

$$f(\mathcal{X}|\mathcal{G}) = \frac{\prod_{e \in E(\mathcal{G})} f(\mathcal{X}_e)}{\prod_{v \in V} f(\mathcal{X}_v)^{d_\mathcal{G}(v)-1}} \qquad (7.3)$$

where $d_\mathcal{G}(v)$ is the degree of v in $\mathcal{G}$. Let BF(e) be the Bayes factor for independence along edge $e = (u, v)$, so that

$$\mathrm{BF}(e) = \frac{f(\mathcal{X}_e)}{f(\mathcal{X}_u)f(\mathcal{X}_v)}.$$

Then (7.3) can be written as

$$f(\mathcal{X}|\mathcal{G}) = \prod_{v \in V} f(\mathcal{X}_v) \prod_{e \in E(\mathcal{G})} \mathrm{BF}(e) \qquad (7.4)$$

It follows from (7.1) and (7.4) that assuming a uniform prior on Γ we can find the MAP estimate by using a maximum weight spanning tree algorithm, using logarithms to the BF(e) as edge weights.

From (41) in Dawid and Lauritzen (1993) we can derive an expression for BF(e) in terms of ratios of gamma functions:

$$\mathrm{BF}(e) = \frac{\Gamma(\lambda_{..} + n_{..})/\Gamma(\lambda_{..}) \prod_{ij} \Gamma(\lambda_{ij} + n_{ij})/\Gamma(\lambda_{ij})}{\prod_i \Gamma(\lambda_{i.} + n_{i.})/\Gamma(\lambda_{i.}) \prod_j \Gamma(\lambda_{.j} + n_{.j})/\Gamma(\lambda_{.j})}$$

where i and j range over the number of levels of X_u and X_v, $\{n_{ij}\}$ is the corresponding table of counts, $\{\lambda_{ij}\}$ the corresponding array of smoothing parameters, and the $\cdot$ notation indicates marginal totals.

The following example illustrates application of this approach to find the MAP forest for a dataset with 400 discrete variables (SNPs). A convenient choice is $\lambda(i) = \alpha/|\mathcal{I}| \; \forall i$, where α is a scalar. This implies that $\lambda_{u,v} = \alpha/(|X_u||X_v|)$ where $|X_u|$ and $|X_v|$ are the number of levels of X_u and X_v. In the following fragment we extract the dataset from the **GGtools** package, define a function to calculate the logarithms of the Bayes factors, and call the `minForest()` function specifying that `logBF` be used to calculate the edge weights. For comparison purposes we also find the minimum BIC forest:

```
> data(hmceuB36.2021)
> p <- 400
> SNPdata <- data.frame(as(smList(MAFfilter(hmceuB36.2021, lower=.1))
+       [["21"]][,1:p],"character"))
> SNPdata[,1:p] <- lapply(SNPdata[,1:p], factor)
> logBF <- function(newEdge, numCat, dataset, alpha=1) {
+    i <- newEdge[1]; j <- newEdge[2]
+    n <- table(dataset[,i], dataset[,j])
+    I <- dim(n)[1];  J <- dim(n)[2]; IJ<-I*J
+    nm <- addmargins(n)
+    ni <- nm[1:I,J+1]; nj <- nm[I+1,1:J]; N <- nm[I+1,J+1]
+    fij <- sum(lgamma(n+alpha/IJ)-lgamma(alpha/IJ))
+    fi <- sum(lgamma(ni+alpha/I)-lgamma(alpha/I))
+    fj <- sum(lgamma(nj+alpha/J)-lgamma(alpha/J))
+    f <- lgamma(N+alpha) - lgamma(alpha)
+    logBF <- fij - fi - fj + f
+    return(logBF)
+ }
> snp.MAP <- minForest(SNPdata, stat=logBF, alpha=1)
> snp.F <- minForest(SNPdata)
```

Then we display the two graphs:

```
> plot(snp.MAP, numIt=500, vert.labels=1:snp.MAP@p, main="MAP forest")
```

MAP forest

```
> plot(snp.F, numIt=500, vert.labels=1:snp.F@p, main="min BIC forest")
```

min BIC forest

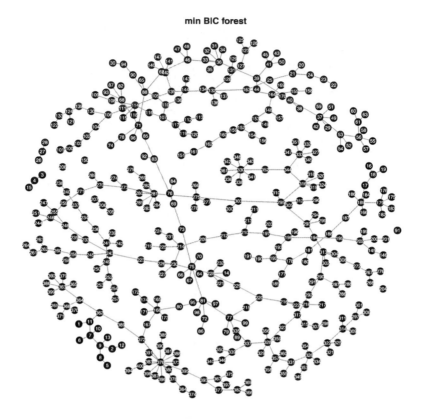

We see that the MAP estimate has many isolated vertices, indicating a stronger tendency to negative logBF values than negative BIC values for weakly associated variables.

References

Akaike H (1974) A new look at the statistical identification problem. IEEE Trans Autom Control 19:716–723

Albert J (2009) Bayesian computation with R, 2nd edn. Springer, New York

Andersson SA, Madigan D, Perlman MD (1996) A characterization of Markov equivalence classes for acyclic digraphs. Ann Stat 25:505–541

Badsberg JH (1991) A guide to CoCo. Tech rep. R-91-43. Department of Mathematics and Computer Science, Aalborg University

Bishop YMM, Fienberg SE, Holland PW (1975) Discrete multivariate analysis: theory and practice. MIT Press, Cambridge

Bøttcher SG, Dethlefsen C (2003) deal: A package for learning Bayesian networks. J Stat Softw 8(20):1–40. http://www.jstatsoft.org/v08/i20

Busk H, Olsen EV, Brøndum J (1999) Determination of lean meat in pig carcasses with the Autofom classification system. Meat Sci 52:307–314

Chickering DM (1996) Learning Bayesian networks is NP-complete. In: Fisher D, Lenz HJ (eds) Learning from data: artificial intelligence and statistics V. Springer, New York, pp 121–130

Chickering DM (2002) Equivalence classes of Bayesian network structure. J Mach Learn Res 2:445–498

Chow CK, Liu CN (1968) Approximating discrete probability distributions with dependence trees. IEEE Trans Inf Theory 14:462–467

Cowell RG, Dawid AP, Lauritzen SL, Spiegelhalter DJ (1999) Probabilistic networks and expert systems. Springer, New York

Dawid AP (1992) Applications of a general propagation algorithm for probabilistic expert systems. Stat Comput 2:25–36

Dawid AP (1998) Conditional independence. In: Kotz S, Read CB, Banks DL (eds) Encyclopedia of statistical sciences, update, vol 2. Wiley-Interscience, New York, pp 146–155

Dawid AP, Lauritzen SL (1993) Hyper Markov laws in the statistical analysis of decomposable graphical models. Ann Stat 21:1272–1317

Dempster AP (1972) Covariance selection. Biometrics 28:157–175

Drton M, Perlman MD (2007) Multiple testing and error control in Gaussian graphical model selection. Stat Sci 22:430–449

Drton M, Perlman MD (2008) A SINful approach to Gaussian graphical model selection. J Stat Plan Inference 138:1179–1200

Edwards D (2000) Introduction to graphical modelling, 2nd edn. Springer, New York

Edwards D, de Abreu GCG, Labouriau R (2010) Selecting high-dimensional mixed graphical models using minimal AIC or BIC forests. BMC Bioinform 11:18

Friedman J, Hastie T, Tibshirani R (2008) Sparse inverse covariance estimation with the graphical lasso. Biostatistics 9(3):432–441

Fruchterman T, Reingold EM (1991) Graph drawing by force-directed placement. Softw Pract Exp 21:1129–1164

Frydenberg M (1990a) The chain graph Markov property. Scand J Stat 17:333–353

Frydenberg M (1990b) Marginalization and collapsibility in graphical interaction models. Ann Stat 18:790–805

Gilks WR, Thomas A, Spiegelhalter DJ (1994) BUGS: a language and program for complex Bayesian modelling. Statistician 43:169–178

Green PJ (2005) GRAPPA: R functions for probability propagation. http://www.stats.bris.ac.uk/~peter/Grappa/

Grizzle JE, Starmer CF, Koch GG (1969) Analysis of categorical data by linear models. Biometrics 25(3):489–504

Højsgaard S, Thiesson B (1995) BIFROST—block recursive models induced from relevant knowledge, observations and statistical techniques. Comput Stat Data Anal 19:155–175

Holm S (1979) A simple sequentially rejective multiple test procedure. Scand J Stat 6:65–70

Johnson RW (1996) Fitting percentage of body fat to simple body measurements. Journal of Statistics Education 4:1

Kirshner S, Smyth P, Robertson AW (2004) Conditional Chow-Liu tree structures for modeling discrete-valued vector time series. In: Proceedings of the 20th conference on uncertainty in artificial intelligence, UAI '04, AUAI Press, Arlington, pp 317–324. http://portal.acm.org/citation.cfm?id=1036843.1036882

Kjærulff U (1990) Graph triangulation—algorithms giving small total state space. Technical report R 90-09, Aalborg University, Denmark

Kruskal J (1956) On the shortest spanning subtree of a graph and the traveling Salesman problem. Proc Am Math Soc 7:48–50

Lauridsen C, Danielsen V (2004) Lactational dietary fat levels and sources influence milk composition and performance of sows and their progeny. Livestock Product Sci 91:95–105

Lauritzen SL (1996) Graphical models. Clarendon Press, Oxford

Lauritzen SL, Spiegelhalter DJ (1988) Local computations with probabilities on graphical structures and their application to expert systems (with discussion). J R Stat Soc B 50:157–224

Ma Z, Xie X, Geng Z (2008) Structural learning of chain graphs via decomposition. J Mach Learn Res 9:2847–2880

Martin PGP, Guillou H, Lasserre F, Déjean S, Lan A, Pascussi JM, Sancristobal M, Legrand P, Besse P, Pineau T (2007) Novel aspects of pparalpha-mediated regulation of lipid and xenobiotic metabolism revealed through a nutrigenomic study. Hepatology 45(3):767–777. http://dx.doi.org/10.1002/hep.21510

Miller LD, Smeds J, George J, Vega VB, Vergara L, Ploner A, Pawitan Y, Hall P, Klaar S, Liu ET, Bergh J (2005) An expression signature for p 53 status in human breast cancer predicts mutation status, transcriptional effects, and patient survival. Proc Natl Acad Sci USA 102(38):13550–13555. http://dx.doi.org/10.1073/pnas.0506230102

Pearl J (2000) Causality. Cambridge University Press, Cambridge

Reiniš Z, Pokorný J, Bazika V, Tišerová J, Goričan K, Horáková D, Stuchlíková E, Havránek T, Hrabovský F (1981) Prognostický význam rizikového profilu v prevenci ischemické choroby srdce. Bratisl Lek Listy 76:137–150

Ripley BD (1996) Pattern recognition and neural networks. Cambridge University Press, Cambridge

Schoener TW (1968) The Anolis lizards of Bimini: resource partitioning in a complex fauna. Ecology 49:704–726

Schwarz G (1978) Estimating the dimension of a model. Ann Math Stat 6:461–464

Sidak Z (1967) Rectangular confidence regions for the means of multivariate normal distributions. J Am Stat Assoc 62(318):626–633. http://www.jstor.org/stable/2283989

Speed TP, Kiiveri H (1986) Gaussian Markov distributions over finite graphs. Ann Math Stat 14:138–150

Spiegelhalter D, Thomas A, Best N, Lunn D (2003) WinBUGS user manual version 1.4. http://www.mrc-bsu.cam.ac.uk/bugs/winbugs/manual14.pdf

Spiegelhalter D, Thomas A, Best N, Lunn D (2011) OpenBUGS user manual version 3.21. http://www.openbugs.info/

Spiegelhalter DJ, Lauritzen SL (1990) Sequential updating of conditional probabilities on directed graphical structures. Networks 20:579–605

Spirtes P, Glymour C (1991) An algorithm for fast recovery of sparse causal graphs. Soc Sci Comput Rev 9(1):62–72. http://ssc.sagepub.com/content/9/1/62.abstract

Spirtes P, Glymour C, Scheines R (1993) Causation, prediction and search. Springer, New York, reprinted by MIT Press

Sturtz S, Ligges U, Gelman A (2005) R2WinBUGS: A package for running WinBUGS from R. J Stat Softw 12(3):1–16. http://www.jstatsoft.org

Tsamardinos I, Aliferis C, Statnikov A (2003) Algorithms for large scale Markov blanket discovery. In: Proceedings of the sixteenth international Florida artificial intelligence research society conference

Verma T, Pearl J (1990) Equivalence and synthesis of causal models. In: Bonissone PP, Henrion M, Kanal LN, Lemmer JF (eds) Uncertainty in artificial intelligence, vol. 6. North-Holland, Amsterdam, pp 255–268

Whittaker J (1990) Graphical models in applied multivariate statistics. Wiley, Chichester

Index

Printed by Publishers' Graphics LLC